GETTING STARTED
WITH
MAPLE®

(For Release 3, 4, and 5)

C-K. Cheung
Boston College

G. E. Keough
Boston College

Michael May, S.J
St. Louis University

John Wiley & Sons, Inc.
New York • Chichester • Weinheim • Brisbane • Singapore • Toronto

Cover image by Marjory Dressler

ISBN 0-471-25249-2

Printed in the United States of America

10 9 8 7 6 5 4 3 2 1

Printed and bound by Hamilton Printing Company
Cover printed by The Lehigh Press, Inc.

To the memory of my Grandmoms

C-K.C.

For Mom and Dad

G.E.K.

For my Parents

M.M., S.J.

Preface

Using this Guide

The purpose of this guide is to give a quick introduction on how to use *Maple V*. It covers Releases 3, 4, and 5 of the software. Also, throughout this guide, we will be suggesting tips and diagnosing common problems that beginners are likely to make. These will make the learning process smoother.

This guide is designed as a self-study tutorial to learn *Maple*. Our emphasis is on getting you quickly "up to speed." This guide can also be used as a supplement (or reference) for students taking a mathematics (or science) course that requires use of *Maple*, such as Calculus, Multivariable Calculus, Advanced Calculus, Linear Algebra, Discrete Mathematics, Modeling, or Statistics.

About *Maple*

Maple is computer algebra software developed by Waterloo Maple Software, which lets you use the computer like an interactive mathematics scratchpad. *Maple* can perform abstract computation as well as numerical computation, graphics, programming and so on. It is a useful tool not only for an undergraduate mathematics or science major, but also for graduate students and researchers. The program is widely used as well by engineers, physicists, transportation officials, and architects.

Organization of the Guide

The Guide is organized as follows:

- Chapter 1 gives a short demonstration of what you'll see in the remaining parts of the Guide. Chapters 2 through 10 contain the basic information that almost every user of *Maple* should know.

- Chapters 11 through 14 demonstrate *Maple*'s capabilities for single-variable calculus. This includes working with derivatives, integrals, series and differential equations.

- Chapters 15 through 20 cover topics of multivariable calculus. Here you'll find discussion of partial derivatives, multiple integrals, vectors, vector fields and line and surface integrals.

- Chapters 21 and 22 introduce the statistical capabilities of *Maple*.

- Chapters 23 through 26 address a collection of topics ranging from animation and simulation to programming and list processing.

- Two appendices explain how to work in the *Maple* worksheet environment and use certain new features of Release 5.

Chapter Structure

Each chapter of the Guide has been structured around an area of undergraduate mathematics. Each moves quickly to define relevant commands, address their syntax, and provide basic examples.

Every chapter ends with as many as three special sections that can be passed over during your first reading. However, these sections will provide valuable support when you start asking questions and looking for more detail. These three sections are:

- "More Examples." Here, you'll find more technical examples or items that address more mathematical points.
- "Useful Tips." This section contains some simple pointers that all *Maple* users eventually learn. We've drawn them from our experiences in teaching undergraduates how to use *Maple*.
- "Troubleshooting Q & A." Here we present a question and answer dialogue on common problems and error messages. You'll also be able to find out *how* certain commands work or *what they assume* you know in using them. Many useful topics are covered here.

Conventions Used in This Guide

Almost every *Maple* input and output you see in this guide appears exactly as we executed it in a *Maple* session. Slight changes were made only to enhance page layout or to guarantee better photo-offset production quality for graphics.

Also, our sections on "Useful Tips" are denoted with *light bulbs* –. More light bulbs indicate tips that we think are more important than others. Our scale is 1 to 4 light bulbs, but your wattage may vary.

Final Comments

First, let us thank all of our colleagues and students who have contributed to the completion of this guide. Special thanks are extended to Jenny Baglivo, Nancy Gaff, and Sarah Quebec for their contributions to the statistics pages of the guide. Bill Keane and Sally Shao contributed several suggestions on the penultimate printing of the guide. Bill Zahner was extremely helpful for his reading, rereading, and critiquing of (many versions of) the guide as it developed. He also offered terrific suggestions on presentation and arrangement.

The staff at John Wiley, Inc., have been very supportive of our efforts, as well as patient with our schedule which always seemed to be slightly missing those deadlines we expected to meet. Our sincerest thanks go to Sharon Smith for her enthusiastic response to this project.

Finally, despite our best efforts, it is likely that somewhere in these many pages, an error of either omission or commission awaits you. We sincerely apologize if this is the case and accept full responsibility for any inaccuracies.

We will make available on the World Wide Web a listing of further comments, examples and any inaccuracies that may be found in this Guide. You can locate this information by navigating the main Boston College website hierarchy down to the Department of Mathematics page. Begin your search at:

<http://www.bc.edu>

We will also try to answer any inquiries you may have about the information presented in the Guide. Feel free to share your teaching/learning experience with us. You can reach us through the following e-mail addresses:

ck.cheung@bc.edu,   keough@bc.edu,   maymk@slu.edu

We wish you only the best computing experiences in *Maple*.

C-K, Jerry, and Mike

Contents

Part I. Basic *Maple* Commands

Part II. Drawing Pictures in *Maple*

Part III. *Maple* for One Variable Calculus

Part IV. *Maple* for Multivariable Calculus

Part V. *Maple* for Linear Algebra and Vector Calculus

Part VI. Using *Maple* in Statistics

CHAPTER 1
Running Maple

Computer Systems

What Computer System Are You Using?

Maple software runs on almost every major computer system including mainframes and desktop systems. *Maple* can also be set up to run across a network and even between systems.

Implementations of *Maple* generally break down into one of two types:

- Text-based systems (e.g., UNIX or other mainframe systems, and DOS-based personal computers) where you can type input through a keyboard only one line at a time.

- Graphically based systems (e.g., desktop systems running MS Windows, Solaris, or the MacOS) where you can both type on a keyboard and use a mouse to navigate a window.

The *Maple* commands we discuss in this guide will work on both types of systems, although some features may apply only to graphically based systems.

Starting the Software

You need to follow the instructions that came with the *Maple* software to install it on your computer system. Once you've completed the installation, you're ready to explore *Maple*.

Starting *Maple* obviously depends on the system you are using:

- On text-based systems, you will enter the command **maple** or a local equivalent, depending on how the software has been installed and how the system is configured. The current version of *Maple* will be launched, and you will see about five lines of text that include a maple leaf and information about the version of *Maple* that has been launched. You will also see the *Maple* prompt, which is usually a greater-than sign (>).

- On graphically based systems, you will typically find the icon of the *Maple* application in a window. Click (or double-click) on the icon. The current version of *Maple* will be launched, and a new window will be opened with the cursor flashing, waiting for your input.

(If you run *Maple* over a network, you may need to check with your system manager for the starting procedure.)

If you get an error message when starting *Maple*, check the Q & A section at the end of this chapter.

Input and Output

Maple is interactive software. For almost every entry you make, *Maple* will provide a direct response. Once you launch *Maple*, you can type, for example:

1 + 1;

and then press the evaluation key or keys (e.g., depending on the release of *Maple* you are using, the **Enter** or **Return** key on a Macintosh/PC, or **Return** on a Sun workstation). *Maple* will give you the response:

> 1 + 1;

>   2

Notes

(1) You end a *Maple* input with a semicolon. If you don't want to see the output, you can end a command with a colon instead.

(2) *Maple* will give you the prompt symbol, >, automatically. You do not need to enter it yourself.

(If you get an error message here, check the Q & A section.)

Now you enter, say,

39 - 11;

then press the **Enter** or **Return** key (depending on your system) and you get:

> 39 – 11;

>   28

Throughout the rest of this manual, we will not show you the prompt sign >. Our input will be shown in boldface and the response in plain text. So, in the future, you will see us write the evaluations above as:

1 + 1;
                2

39 - 11;
                28

Quit

If you've had enough and want to exit *Maple*, you can simply enter:

quit;

On graphically based systems, you can finish a *Maple* session by choosing the **Quit** item in the File menu.

A Quick Tour

For the rest of this chapter, we'll show you some of *Maple*'s capabilities. We present these examples only to whet your appetite. You can follow along at your computer by typing what we show below. In later chapters, we'll give you a more complete explanation of how to use these commands.

Note: When you input the following commands in *Maple*, make sure that:

- You use upper- and lower-case characters exactly as we do. *Maple* is very "case sensitive." If you use the wrong capitalization, you can hurt *Maple*'s "feelings."

- You use exactly the type of brackets we show. There are three types of brackets: [*square brackets*], (*parentheses*) and { *curly braces* }. Each has its own meaning in *Maple*. If you use the wrong one, *Maple* will be confused.

- On a text-based system, if a command is more than one line, you can continue to a second line by pressing **Return** without using a semicolon, but you should break the line exactly as we show in the text. On a graphically based system, either **Shift Return** or **Return**, depending on your system, lets you continue the command on a second line without evaluating it right away.

- You press the "evaluation key" (e.g., the **Enter** key on a Macintosh or the **Shift Enter** combination on a PC) to see the output.

Calculator

Maple does all the work of a hand-held electronic calculator. You can enter numerical expressions and *Maple* will do the arithmetic:

```
235.567*441.235/623.45;
```
$$166.7181092$$

```
sin(0.3);
```
$$.2955202067$$

But *Maple* can go much further. Try this factorial computation!

```
289!;
```
2079866075306145164348895732262527092227125189083652864966524223174057 6\
0295930638776430109826354519132675660433931363055910963871453772379754 9\
3144476665273919230320176358872361834759374038542872584612257227104981 8\
9168763234932439760233029166663945402474493070106657313319035568964279 6\
2603583329193202835131888878612868953848908671300544998959169558544601 48\
0588131077161074357869689701962388257295731157260337104837625533823057 2\
5385845880790786699431748508548589955805949142445628564109186070285204 1\
0684463210865874698240000000000000000000000000000000000000000000000000 0\
00000000000000000000

You can't get this on your calculator!

Solving Equations

Maple can solve complicated equations and even systems of equations in many variables. For example, the equations $2x + 5y = 37$ and $x - 3y = 21$ have a simultaneous solution:

```
solve({ 2*x+5*y = 37, x-3*y = 21 }, { x, y });
```
$$\left\{ y = \frac{-5}{11}, \quad x = \frac{216}{11} \right\}$$

Maple can also find solutions to equations numerically. For example, the equation $x = \cos(x)$ has a solution very close to $x = 0.75$. We can find it with:

```
fsolve( x=cos(x), x );
```
$$.7390851332$$

Computer Algebra

Maple is very good at algebra. It can work with polynomials:

```
expand((x-2)^2 * (x+5)^3);
```

$$x^5 + 11x^4 + 19x^3 - 115x^2 - 200x + 500$$

```
factor(x^5+11*x^4+19*x^3-115*x^2-200*x+500);
```

$$(x-2)^2(x+5)^3$$

Are you impressed? *Maple* also knows standard trigonometric identities such as $\sin^2(x) + \cos^2(x) = 1$:

```
simplify(sin(x)^2+cos(x)^2);
```

$$1$$

That one was easy, but you probably forgot that $\sec^2(x) - \tan^2(x) = 1$:

```
simplify(sec(x)^2-tan(x)^2);
```

$$1$$

Calculus

Maple even knows a lot about calculus! We can find the derivative of the function $f(x) = x/(1+x^2)$ with:

```
diff(x/(1+x^2),x);
```

$$\frac{1}{1+x^2} - 2\frac{x^2}{(1+x^2)^2}$$

A complicated integral such as $\int \frac{1}{1+x^3}\, dx$ is handled rather easily.

```
int(1/(1+x^3),x);
```

$$\frac{1}{3}\ln(1+x) - \frac{1}{6}\ln(x^2 - x + 1) + \frac{1}{3}\sqrt{3}\arctan\left(\frac{1}{3}(2x-1)\sqrt{3}\right)$$

Graphing in the Plane

Maple does everything a standard graphing calculator does and does it better! For example, to see the graph of the function $f(x) = x/(1+x^2)$ over the interval $-4 \le x \le 4$, you can use:

```
plot( x/(1+x^2), x=-4..4 );
```

We can see a "daisy" with:

```
plot([cos(21*t)*cos(t), cos(21*t)*sin(t), t=0..2*Pi]);
```

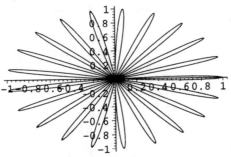

Plotting in Space

Maple does a wonderful job with three-dimensional graphics. Let us show you some pictures.

```
plot3d(sin(x)*cos(y), x=0..2*Pi, y=0..2*Pi);
```

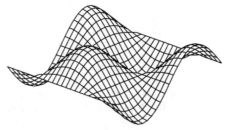

```
plot3d([t/5, r*cos(t), r*sin(t)], r=0..1, t=0..6*Pi,
       grid=[8,60]);
```

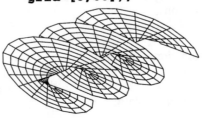

Changing the 3-D View

If you are using *Maple V* Release 5, you can click and hold the (left) mouse button with the cursor/arrow at any point inside a 3-D picture. As you drag the mouse with the button held down, the picture rotates. This allows you to see the 3-D picture from any viewpoint.

Programming

You can program *Maple* much as you do when you are using a programming language such as Pascal or C. Here's a simple routine to simulate the flipping of a coin several times and return the number of heads observed:

```
CoinFlips := proc(howmany)
        local heads, counter, toss;
        heads := 0:
        toss := rand(0..1);
        for counter from 1 to howmany do
          if toss() = 1 then heads := heads + 1 fi;
        od;
        heads;
        end:
```

You can use this routine with statements such as:

print(CoinFlips(100),` heads seen in 100 trials.`);

55 heads seen in 100 trials.

print(CoinFlips(1000),` heads seen in 1000 flips.`);

491 heads seen in 1000 trials.

More Examples

The "More Examples" sections of this guide present examples involving more mathematics. Students of mathematics, science, and engineering may find these of interest.

Here's one example. Not too many people know about the Bessel functions. But if you're learning physics, you might want to know what *Maple* has available for you in Bessel functions.

Special Functions

The Bessel function of order 0, $J_0(x)$, is a solution to the differential equation:

$$x^2 y'' + xy' + x^2 y = 0.$$

You can see its graph with:

plot(BesselJ(0,x),x=0..20);

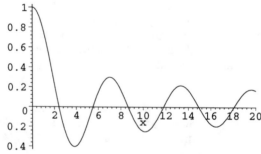

The smallest, positive zero of $J_0(x)$ is at approximately $x = 2.40483$:

fsolve(BesselJ(0,x)=0,x);

2.404825558

Useful Tips

 If you are running *Maple* on your own computer, it is better to quit other applications before you start *Maple*. If you need to run another program at the same time as you run *Maple*, start with *Maple* first.

Troubleshooting Q & A

Most chapters of this guide end with a Troubleshooting section, where we answer some common questions we think you will have. But there's not too much you can ask yet, except:

Question ... When I tried to start *Maple*, I got an error message that there wasn't enough memory. What should I do?

Answer ... *Maple* is memory intensive. Make sure that you have enough RAM to run the software. Also, it is a good practice to quit other applications before you launch *Maple*.

Question ... If I forgot to end a *Maple* command with a semicolon, but hit return/enter anyway, what would happen?

Answer ... Depending on your system, failing to add the semicolon can cause an error message, or the system may simply give a new prompt, >, and wait for you to finish the input.

Question ... What do I do now?

Answer ... Turn the page and start learning about *Maple*! Also, if you are using *Maple V* Release 4 or higher on a graphically based system, check the Appendices for how to use a "worksheet".

CHAPTER 2

Calculator Features in Maple

Simple Arithmetic

Basic Arithmetic Operations

In this chapter, we will show you how *Maple* can work as a calculator. We start with basic arithmetic.

The basic arithmetic operations of *Maple* are:

Command	What It Does
+, -	Add, Subtract
*, /	Multiply, Divide
^	Raise to a power (exponentiation)

Here are some expressions to evaluate: $\dfrac{25.5}{5}$, $4+2^5$ and $\dfrac{23}{5}-\dfrac{3}{5}+5(2^3)$:

```
25.5/5;
```
$$5.100000000$$

```
4+2^5;
```
$$36$$

```
23/5 - 3/5 + 5*2^3;
```
$$44$$

We can also use parentheses to group terms together. For example, the expression $(3+4)\left(\dfrac{4-8}{5}\right)$ is entered with:

```
(3+4)*((4-8)/5);
```
$$-\frac{28}{5}$$

Note: Use (*parentheses*) to group terms in expressions. Do not use [*square brackets*] or { *curly braces* }. They mean something different.

8

Precedence　　*Maple* follows the laws of precedence of multiplication over addition and so on, just as you do by hand. For example,

9/6*22+5;

　　　38

actually computes $(\frac{9}{6} \times 22) + 5$. A common mistake is to think of the input as either $\frac{9}{6 \times 22} + 5$ or $\frac{9}{6} \times (22 + 5)$.

Comments　　You can add a comment to any expression by starting the comment with the pound sign #. For example:

27*3;　　　　　　　# This multiplies 27 and 3.

　　　81

Maple will neglect the phrase "**This multiplies 27 and 3.**" It is for your own reference.

We'll use comment lines throughout this guide to write short reminders about what we're emphasizing in certain examples.

Previous Results and " Syntax

As a convenience, *Maple* (Release 4 or earlier) lets you use the quotation sign, ", to stand for "the last result obtained." In this way, you can avoid retyping output in your next input.

For example,

3200*12;

　　　38400

" - 6500;　　　# Same as 38400 - 6500.

　　　31900

(1000 + ")^2;　# Same as (1000+31900)^2.

　　　1082410000

You can use two quotation signs "" for the "second-last result obtained," and three quotation signs """ for the third-last.

In Release 5, *Maple* now uses the percentage sign, %, instead of ". Thus the sequence above would appear in Release 5 as:

3200*12;

　　　38400

% - 6500;　　　# Same as 38400 - 6500 in Release 5.

　　　31900

(1000 + %)^2;　# Same as (1000+31900)^2 in Release 5.

　　　1082410000

> **Note:** Using the quotation or percentage sign can be confusing. We will therefore use it sparingly.

Output Styles

Calculator-Style Values

Maple gives you an exact (symbolic) value for every expression:

        (3+9) * (4-8) / 1247 * 67;

$$\frac{-3216}{1247}$$

You can force *Maple* to give you an answer that looks like the decimal answer you'd get on a calculator by using **evalf** (for floating point evaluation) with parentheses around an expression:

        evalf((3+9) * (4-8) / 1247 * 67);
              -2.578989575

You can see more digits in the answer – say 40 – with:

        evalf((3+9) * (4-8) / 1247 * 67 , 40);
        $- 2.578989574979951884522854851643945469126$

Results like these are called **approximate numeric values** in *Maple*.

Scientific Notation

Maple uses a modified standard scientific notation to display results when the numbers either get very large or very small:

        evalf(1234567890);

              $.1234567890 \ 10^{10}$

        0.000003492836 ;

              $.3492836 \ 10^{-5}$

The spaces in the output above represent multiplication.

Built-in Constants and Functions

Built-in Constants

The mathematical constants used most often are already built into *Maple*. Be careful using upper-case and lower-case characters when you use these constants.

Constant	Value	Explanation	Maple
π	3.1415926...	Ratio of a circle's circumference to its diameter	**Pi**
e	2.71828...	Natural exponential	**exp(1)**
i	$i = \sqrt{-1}$	Imaginary number	**I**
∞	∞	(Positive) infinity	**infinity**

For example, to see the value of π to 45 significant digits, we use:

```
evalf(Pi, 45);
```
$$3.14159265358979323846264338327950288419716940$$

To compute the numerical value of $\pi^4 - 5e^{1/3}$, we type:

```
evalf(Pi^4 - 5*exp(1)^(1/3));
```
$$90.43102896$$

Built-in Functions

Maple has many built-in functions. Here are the functions you will probably use the most.

Function(s)	Sample(s)	Maple Name(s)
Natural logarithm	$\ln(x)$	**ln(x)**
Logarithm to base *a*	$\log_a x$	**log[a](x)**
Exponential	e^x	**exp(x)**
Absolute value	$\lvert x \rvert$	**abs(x)**
Square root	$\sqrt{x}$	**sqrt(x)**
Trigonometric	$\sin(x)$, $\cos(x)$, …	**sin(x) cos(x), tan(x), cot(x), sec(x), csc(x)**
Inverse trigonometric	$\sin^{-1}(x)$, $\cos^{-1}(x)$, …	**arcsin(x), arccos(x), arctan(x), arccot(x),** etc.
Hyperbolic	$\sinh(x)$, $\cosh(x)$, …	**sinh(x), cosh(x), tanh(x), coth(x), sech(x), csch(x)**
Inverse hyperbolic	$\sinh^{-1}(x)$, $\cosh^{-1}(x)$, …	**arcsinh(x), arccosh(x), arctanh(x), arccoth(x),** etc.

For example,

```
sin(Pi);
```
$$0$$

```
evalf(sin(180));      # 180 radians, NOT 180 degrees.
```
$$-.8011526357$$

```
arctan(1);
```
$$\frac{1}{4}\pi$$

```
exp(ln(exp(1)));
```
$$e$$

Notes

(1) *Maple* uses radian measure for all angles.

> (2) *Maple* refers to e^1 as **e** in the output. Unfortunately, you cannot use **e** when you enter input. You have to write **exp(1)** for input.

Error Messages

If you make a mistake in your input, *Maple* will *print an error message* when you evaluate it. In the graphically based versions of *Maple* V Release 4 or higher, you will get an error message printed, and the cursor will move to the place where *Maple* thinks the mistake is located.

Two Common Mistakes

The two most common mistakes for beginners are:

* Mismatching parentheses. For example:

  ```
  3*(4-5))+6;
  Syntax error, `)` unexpected
  ```

  ```
  sin 3;        # It should be sin(3);
  Syntax error, unexpected number
  ```

* Forgetting the multiplication sign. For example:

  ```
  5+2x^2;       # It should be 5+2*x^2;
  Syntax error, missing operator or `;`
  ```

More Examples

Approximate Numbers and Exactness

Working with **approximate numeric values** is just like working with values on a hand-held calculator. These numbers sometimes lose precision as values get rounded off in arithmetic.

■ **Example.** *Maple* makes a big distinction between the exact number π and a numerical approximation for it. For example, **sin(517*Pi)** is an exact quantity with an exact answer:

```
sin(517*Pi);
```

$$0$$

If you use a numerical approximation for 517π, you don't get an exact zero:

```
sin(evalf(517*Pi));
```

$$-.9407689571 \ 10^{-7}$$

This is very close to zero (it is, after all, -0.00000009407689571) and it's probably acceptable for the work you will be doing. But it's not exact.

Useful Tips

💡 💡 💡 💡 Don't forget to type the * when multiplying terms. This is the most common mistake that beginners make. For example, you might incorrectly enter **23cos(0)** instead of **23*cos(0)**.

💡 💡 💡 💡 Never try to multiply terms together in *Maple* using (*parentheses*). For example, in mathematical writing you can write (3)(4) to denote 3×4. However, *Maple* has a different interpretation.

(3)(4);

$$3$$

💡 💡 💡 Use parentheses in expressions to clarify what you mean. This helps avoid mistakes. For example, you might think that **3^2*x** means 3^{2x}, but it doesn't! It is actually $(3^2)x$ because the square is done before the multiplication of **2** and **x**. To get 3^{2x}, you should write **3^(2*x)**.

💡 💡 Avoid using the " or % symbols except for short sequences of one or two calculations.

💡 Avoid using **exp(1)** together with ^ to describe an exponential function. For example, the expression e^{2x} is better written as **exp(2*x)**, rather than **exp(1)^(2*x)**.

Troubleshooting Q & A

Question... When I tried to evaluate a built-in function, *Maple* gave an error message, "`Syntax error, unexpected number`." What should I check?

Answer... Make sure that you included parentheses when using *Maple* functions. For example, a common mistake for beginners is to type:

sin 2;
`Syntax error, unexpected number`

The correct input should be **sin(2);**.

Question... When I entered **sin(2)**, *Maple* just gave me the same thing back again. It didn't evaluate it. Why?

sin(2);

$$\sin(2)$$

Answer... *Maple* always gives you an exact answer. When you write sin(2) in mathematics, you don't try to simplify it. Neither does *Maple*. However, if you want a decimal approximation for sin(2), use **evalf**.

```
evalf(sin(2));
```
.9092974268

Question... When I tried to evaluate a built-in function, *Maple* just returned the input unevaluated. I then used **evalf**, but still *Maple* did not give me a numerical value. What should I check?

Answer... Check your spelling. Most likely, you've misspelled the name of a built-in function or constant. For example, you may have used **cosine(Pi)** or **cos(pi)** instead of **cos(Pi)**.

Question... I got the wrong answer when I entered a complicated arithmetic expression. What should I check?

Answer... Always use (*parentheses*) to keep your expressions manageable and readable. Sometimes you can get tripped up by not knowing the exact order in which expressions are evaluated. Parentheses make the order of evaluation clear.

For example, $(2^3)^4 = 4096$ and $2^{(3^4)}$ are very different numbers:

```
(2^3)^4;
```
4096

```
2^(3^4);
```
2417851639229258349412352

But if you don't use parentheses, you might not know which one gets computed. (In this case *Maple* will not evaluate the ambiguous expression.)

```
2^3^4;        # The order of operations is not clear.
```
Syntax error, `^` unexpected

Question... I got the error message "Syntax error, missing operator or `;`" when I entered a complicated arithmetic expression. What should I check?

Answer... There are many possible errors that will generate this error message. You should:
- Check all the spelling.
- Check if all the (*parentheses*) are in the right places. In particular, make sure you correctly matched right and left parentheses.
- Check that you typed * for multiplication. For example:

```
3sin(5) + 7;
```
Syntax error, missing operator or `;`

```
3*sin(5) + 7;
```

CHAPTER 3
Variables and Functions

Variables

Immediate Assignment

With a calculator, you can store a value into memory and then recall it later. With a more advanced calculator, you can store different values under names such as A, B, C, and so on.

You can do even better in *Maple*. You can assign a name to any *Maple* expression or value and then recall it whenever you want. You do this using a colon followed by an equal sign :=, which is the symbol for **immediate assignment**. For example:

a := 3.4;   # We assign **a** to be the value 3.4.

$$a := 3.4$$

(Note that there is no space between the colon and equal sign.)

Once you've made an assignment, you can recall its value or use it in an expression:

a;

3.4

a + 2;

5.4

a^2;

11.56

How Expressions Are Evaluated

You may want to know how *Maple* keeps track of all the symbols and variables that you have defined.

Say we assign the name **myLunch** to the sum of **apple** and 3 times **banana**.

myLunch := apple + 3 * banana;

$myLunch := apple + 3\ banana$ # *Maple* spits back the definition
                                  # because **apple** and **banana** have
                                  # no associated value.

Now suppose we give the value 2 to **apple** and then ask that **myLunch** be reevaluated:

apple := 2;   # Now **apple** has the value 2.

$apple := 2$

myLunch;

$2 + 3 \ banana$ # When *Maple* reevaluates **myLunch**,
                        # it substitutes the value 2 for **apple**.

If we now define the value of **banana** to be 3 and reevaluate **myLunch**:

banana := 3;   # Now **banana** has the value 3.

$banana := 3$

myLunch;

11   # When *Maple* reevaluates **myLunch**, it replaces
                # **apple** and **banana** by their respective values, and
                # simplifies the resulting expression. Bon appétit!!

Redefining and Clearing Symbols

Once a symbol name has been assigned a value or an expression, *Maple* retains the association until you end your session, redefine it, or explicitly clear it.

a := 3.4;   # We assign **a** to be 3.4.

$a := 3.4$

a := 5;   # We reassign **a** to be 5.

$a := 5$

a + 2;   # *Maple* uses the most recently assigned value of **a**.

7

The easiest way to ask *Maple* to forget about an assignment is to assign the variable back to its own name.

a := 'a';   # Note that we use two single quotes around **a**.

$a := a$

a;   # We can see that **a** is no longer 5.

a

In *Maple V* Release 4 or higher, you can also use the **unassign** command to clear a variable. (In Release 3 or lower, you need to type **readlib(unassign);** before you can use **unassign**.)

unassign('a');   # Note that we use two single quotes around **a**.
a;

a

You can also **unassign** assignments for several names at the same time, using one statement:

unassign('myLunch', 'apple', 'banana');

Rules for Names

Names you use can be made up of letters and numbers, subject to the following two rules:

- You can't use a name that begins with a number. For example, **2app** is not an acceptable name because *Maple* will think that you forgot to type an operator between "2" and "**app**," and so will give you an error message.

- You can't choose names that conflict with *Maple*'s own names. For example, you can't name one of your own variables **sin**.

All of the following are examples of legitimate names that you could use:

a, m, p1, A, area, Perimeter, Batman, good4you, classsOf2001

> **Note:** *Maple* distinguishes upper case and lower case characters. For example, the names **Batman**, **batman**, and **batMan** are different.

Substitution Command

You can substitute values into an expression without defining the variables explicitly. The substitution command, **subs**, is used in the form:

subs (*list of substitutions using =* **,** *expression* **);**

For example, to substitute $x = 2$ and $y = 5$ into the expression $x^2 - 2xy$:

```
subs(x=2, y=5, x^2 - 2*x*y );
```
$$-16$$

One advantage of using the substitution command is that the value you substitute into a variable is temporary and is not remembered by *Maple*.

```
x;                    # The value of x is unchanged by the previous substitution.
```
$$x$$

Finding the Larger Root of a Quadratic Equation

■ **Example**. We want to find the larger root of each of the following quadratic equations: $2x^2 + 5x - 6 = 0$ and $2x^2 + x - 3 = 0$.

The roots of a quadratic equation $ax^2 + bx + c = 0$, with $a \neq 0$, are found using the quadratic formula $(-b \pm \sqrt{b^2 - 4ac})/(2a)$. The larger root, when $a > 0$, is thus:

```
unassign('a', 'b', 'c', 'largerroot');
largerroot := (- b + sqrt(b^2 - 4*a*c) )/(2*a);
```
$$largerroot := \frac{-b + \sqrt{b^2 - 4ac}}{2a}$$

Here are the larger roots of each of the two equations:

```
subs(a=2, b=5, c=-6, largerroot);
```
$$-\frac{5}{4} + \frac{1}{4}\sqrt{73}$$

```
subs(a=2, b=1, c=-3, largerroot);
```
$$\frac{-1}{4} + \frac{1}{4}\sqrt{25}$$

The above answer can actually be simplified to 1. (*Maple* does not notice that $\sqrt{25}$ = 5.) You can tell *Maple* to do so by using the **simplify** command which will be discussed in detail in the next chapter.

```
simplify(");              # or simplify(%); for R5 (Release 5).
```
$$1$$

Functions

Defining Functions

Maple has many built-in functions such as **sqrt**, **sin**, and **tan**. You can add your own functions as well.

To define a function $f(x)$ in *Maple*, you use:

> **f := x -> *formula* ;**

The "arrow symbol" **->** is formed by entering the minus sign and greater-than sign together, with no spaces in between.

For example, you can define the function $f(x) = x^2 + 5x$ in *Maple* with:

> **f := x -> x^2+5*x;**

$$f := x \rightarrow x^2 + 5x$$

Now, we can do some evaluations:

> **f(7.1);**

$$85.91$$

> **f(a);**

$$a^2 + 5a$$

> **f(x+1);**

$$(x+1)^2 + 5x + 5$$

> **f(f(y));**

$$(y^2 + 5y)^2 + 5y^2 + 25y$$

Functions with More Than One Variable

Functions may have more than one variable. A simple example is the computation of the average speed of an automobile.

If an automobile travels m miles in the span of t minutes, then its average speed in miles per hour is given by the expression $\frac{m}{(t/60)} = \frac{60m}{t}$. We then have a speed function "$f(m, t) = \frac{60m}{t}$":

> **speed := (m,t) -> 60*m / t ;**

$$speed := (m, t) \rightarrow 60\,\frac{m}{t}$$

If a distance of 45 miles is traveled in 30 minutes, the average speed will be 90 m.p.h.:

> **speed(45,30);**

$$90$$

More Examples

Functions of Split Definition

Functions sometimes cannot be defined using a single formula. For example, the famous Heaviside function is defined by:

$$H(x) = \begin{cases} 1, & \text{if } x > 0 \\ 0, & \text{if } x \le 0 \end{cases}$$

To define this function in *Maple*, we use the **piecewise** command as follows:

h := x -> piecewise(x > 0, 1, x <= 0, 0);

$$h := x \rightarrow piecewise(x > 0, 1, x \le 0, 0)$$

(The characters "<=" in the command mean $\le$, less than or equal to.)

To use **piecewise** to define a function having two branches, write:

piecewise(*condition1* **,** *result1* **,** *condition2* **,** *result2* **)**

This means that if *condition1* is satisfied, then *result1* will be used; otherwise, *Maple* will use *result2* if *condition2* is satisfied. The following table shows some operators you will use to check conditions.

Operator	*Meaning*	*Operator*	*Meaning*
=	Equal to	**<>**	Not equal to
>	Greater than	**<**	Less than
>=	Greater than or equal to	**<=**	Less than or equal to
and	And	**or**	Or

The syntax for the **piecewise** command can also be extended. For example, the following function *f* has three branches and can be defined as you see below:

$$f(x) = \begin{cases} 1 - x, & \text{if } 1 < x < 3 \\ x^2, & \text{if } 0 \le x \le 1 \\ x + 2, & \text{if } x < 0 \text{ or } x \ge 3 \end{cases}$$

f := x -> piecewise(1 < x and x < 3, 1-x,
**   0 <= x and x <= 1, x^2, x < 0 or x >= 3, x+2);**

To see this definition more clearly, we recommend that you write the conditions one on each line and line them up carefully:

f := x -> piecewise(1 < x and x < 3,   1-x,
**   0 <= x and x <= 1,   x^2,**
**   x < 0 or x >= 3,   x+2);**

Useful Tips

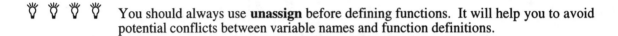 Never assign values to any of the names **x**, **y**, **z**, or **t**. Otherwise, *Maple* will confuse them with the variables **x**, **y**, **z**, or **t** that you typically use when defining functions.

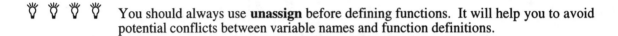 You should always use **unassign** before defining functions. It will help you to avoid potential conflicts between variable names and function definitions.

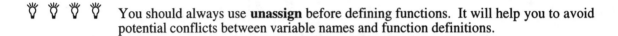 Most of *Maple*'s built-in names use only lower case letters. If the names of your own variables and functions contain a capital letter, (e.g. **myFunction**, **newVar**), you will be able to distinguish them from *Maple*'s.

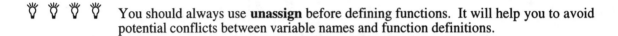 We recommend that you define **e := exp(1);**. This can simplify many inputs. For example, $e^{0.3}$ can now be entered as **e^0.3**, instead of **exp(1)^0.3**.

Troubleshooting Q & A

Question... When I defined a new variable or function, I got the error message "`attempting to assign to ... which is protected.`" What did I do wrong?

Answer... When you try to rename a built-in function or built-in constant, *Maple* will give you this error message. You cannot choose names that conflict with *Maple*'s own names.

Question... I tried to define a function, but I couldn't get it to work. What should I check?

Answer... Three things usually bother function definitions.

- First, check for the proper syntax. Make sure you use a colon-equal definition. Also, don't leave any space between the – and > keys in forming the –> sign. (Common mistakes include using = instead of **:=** and using **f(x) :=** instead of **f := x ->**.)

- Second, check that you have entered the formula of the function correctly. Some common mistakes are to use expressions such as **2x**, **cos x**, **e^x**, and so on.

- Third, your function or variable name may conflict with something else you used earlier in your *Maple* session. For example, you may have once defined:

  x := 3;

  Later you define:

  f := x -> x^2;
  f(x);

  　　　　　　9

The result is not the x^2 you expected, but $3^2 = 9$ since **x** has the value 3.

Make sure you explicitly clear the variable(s) before you define your function:

```
x := 'x';
f := x -> x^2;
```

Please note that even if you start a new worksheet in a *single Maple* session, all the variables or functions that you defined in earlier worksheets will still be retained until you quit *Maple*.

Question... I'm still not sure about when I should use an equal sign = or when I should use a colon-equal := definition?

Answer... The symbol = means equality, while := means assignment. Equality is a test that gives a true or false answer. Assignment is an action that either gives a name to a value or defines a function.

Question... When I tried to evaluate a function, I got the error message "... uses a 2nd argument, ... , which is missing." What went wrong?

Answer... This message means that you did not supply enough variables to evaluate the function.

A common mistake is to define a function of one variable, say,

```
f := x -> x^2;
```

Later, you use the same name **f** to define a function of two variables, say,

```
f := (x, y) -> x + 2*y;
```

The original definition of $f(x) = x^2$ has now been erased. If you type **f(x)**, you will get an error message complaining about not having enough variables (because *Maple* is now expecting two variables for the function).

Question... When I used **unassign**, I got the error message "... invalid arguments." What went wrong?

Answer... Make sure that you use two single quotes around the variable or function name that you want to clear. For example:

```
unassign(a);        # This will give you an error message.
```

```
Error, (in assign) invalid arguments
```

```
unassign('a');      # This is correct.
```

CHAPTER 4
Computer Algebra

Working with Polynomials and Powers

In this chapter we will show you how common algebraic operations can be done directly in *Maple*. Let's start with polynomials.

The expand and factor Commands

The **expand** command does exactly what its name says it does:

```
expand((x-2)*(x-3)*(x+1)^2);
```

$$x^4 - 3x^3 - 3x^2 + 7x + 6$$

The **factor** command is basically the reverse of the **expand** command:

```
factor(x^4-3*x^3-3*x^2+7*x+6);
```

$$(x-2)\ (x-3)\ (x+1)^2$$

Here are some more examples:

Example	Comment
`factor(x^2-3);` $x^2 - 3$	Although $x^2 - 3 = (x + \sqrt{3})(x - \sqrt{3})$, **factor** will not give radicals in its answer.
`factor(x^2-3,sqrt(3));` $(x + \sqrt{3})\ (x - \sqrt{3})$	We can give *Maple* a list of radicals that it is to allow to use in factorization.
`expand((x-1.54)*(3.2*x-2.9));` $3.2x^2 - 7.828x + 4.466$ `factor(");` # or **factor(%);** for R5. $3.2\ (x - .90625000)\ (x - 1.5400000)$	**factor** does a nice job even when you use numerical coefficients.

`factor((5-3*I)+(-4+I)*x +`
`    (1-I)*x^2);`
 	
$\left(\dfrac{1}{2} - \dfrac{1}{2}I\right)(2x - 3 - 5I)\ (x - 1 + I)$
	factor works even when the coefficients are complex.

`factor(x^2+1);` $x^2 + 1$	**factor** won't use complex numbers unless at least one of the coefficients is a complex number.
`factor(x^2+1, I);` $(x + I)\ (x - I)$	You can force a factoring into complex terms by listing **I** as the optional second input.

```
expand((x-y+z)^3);
```

$$x^3 - 3x^2y + 3x^2z + 3xy^2 - 6xyz +$$
$$3xz^2 - y^3 + 3y^2z - 3yz^2 + z^3$$

```
factor(");          # or factor(%) for R5.
```

$$(x - y + z)^3$$

The **expand** and **factor** commands can also work for polynomials with more than one variable.

The simplify Command

The **simplify** command tries to produce an expression that *Maple* thinks is the simplest form. Most times, this will be the same as what you think of as "simplest."

Consider, for example, $(x+1)^2 - 4x = x^2 - 2x + 1 = (x-1)^2$. The term on the left-hand side would not be considered simplified, and neither would the middle term. Let's see how the **simplify** command reacts to them.

```
simplify((x+1)^2-4*x);
```

$$x^2 - 2x + 1 \qquad \text{\# \textit{Maple} thinks that this is the simplest form.}$$

```
simplify(x^2-2*x+1);
```

$$x^2 - 2x + 1 \qquad \text{\# \textit{Maple} thinks that this is the simplest form already.}$$

```
simplify((x-1)^2);
```

$$(x-1)^2 \qquad \text{\# \textit{Maple} thinks that this is the simplest form too!}$$

The symbolic Option

By default, *Maple* is very careful in simplifying. It understands that **sqrt(x^2)** is not always the same as **x** for all complex numbers. Instead, **sqrt(x^2)** simplifies to **x** times the complex sign of **x**.

```
simplify(sqrt(x^2));
```

$$\operatorname{csgn}(x)\,x \qquad \text{\# csgn stands for complex sign.}$$

Sometimes we want such expressions of powers simplified anyway. In those cases we use the **symbolic** option of the **simplify** command.

```
simplify(sqrt(x^2), symbolic);
```

$$x$$

The **symbolic** option of the **simplify** command will handle $\sqrt{x^2}$ by treating it symbolically as $(x^2)^{1/2}$ and rewriting it to be $x^{2(1/2)} = x^1$.

Similarly:

```
simplify((x^6)^(1/3));
```

$$\left(x^6\right)^{1/3}$$

```
simplify((x^6)^(1/3), symbolic);
```

$$x^2$$

Working with Rational Functions

The simplify and convert Commands

A rational function is an expression of the form $\frac{a\ polynomial}{another\ polynomial}$. The following are three common algebraic operations involving rational functions.

- Combining terms over a common denominator can be done with the **simplify** command. For example, to combine $\frac{2}{3x+1} + \frac{5x}{x+2}$:

```
simplify( 2/(3*x+1) + (5*x)/(x+2));
```

$$\frac{7x+4+15x^2}{(3x+1)\ (x+2)}$$

- Splitting up rational functions into partial fractions can be done with the **convert** command if you specify the **parfrac** option. For example, to split up $\frac{11x^2 - 17x}{(x-1)^2(2x+1)}$:

```
convert((11*x^2-17*x)/((x-1)^2*(2*x+1)),parfrac,x);
```

$$-\frac{2}{(x-1)^2} + \frac{3}{(x-1)} + \frac{5}{(2x+1)}$$

- The **convert** command also does long division when you use the **parfrac** option. For example, $(x^5 - 2x^2 + 6x + 1) \div (x^2 + x + 1)$ can be found with:

```
convert((x^5-2*x^2+6*x+1)/(x^2+x+1),parfrac,x);
```

$$x^3 - x^2 - 1 + \frac{2+7x}{x^2+x+1}$$

Working with Trig and Hyperbolic Functions

The **simplify**, **expand**, and **factor** commands can also be used for expressions that involve trigonometric functions, but their results may not come out the way you think they should.

Basic Trig and Hyperbolic Identities

The **simplify** command can recognize many basic trigonometric or hyperbolic identities:

```
simplify(sin(x)^2+cos(x)^2);
```

$$1$$

```
simplify(sin(x)^2-cos(x)^2);
```

$$-2\cos(x)^2 + 1$$

```
simplify(cosh(x)^2-sinh(x)^2);
```
$$1$$

The **expand** command works on trig and hyperbolic functions without any difficulties.

```
expand(sin(2*x));
```
$$2\sin(x)\cos(x)$$

```
expand(cosh(2*x));
```
$$2\cosh(x)^2 - 1$$

The combine Command

The **factor** command is not very effective in working with trig and hyperbolic functions. For example,

```
factor(sin(x)^2-cos(x)^2);
```
$$(\sin(x) - \cos(x))\ (\sin(x) + \cos(x))$$

Instead we need to use the **combine** command to produce better results.

```
combine(sin(x)^2-cos(x)^2);
```
$$-\cos(2x)$$

```
combine(2*sin(x)*cos(x));
```
$$\sin(2x)$$

```
combine(cosh(x)^2+sinh(x)^2);
```
$$\cosh(2x)$$

```
combine(sin(2*x)*cos(3*x));
```
$$\frac{1}{2}\sin(5x) - \frac{1}{2}\sin(x)$$

Useful Tips

 You have to be careful when you use the **symbolic** option in the **simplify** command. Most of the time you use it, you will be assuming that all the quantities you're working with are non negative real numbers.

Troubleshooting Q & A

Question... When I used **simplify**, **expand**, or **factor** on a polynomial, I got a number. What went wrong?

Answer...   First, it is possible that after expansion or simplification, all the terms in your polynomial canceled out and left you with a constant.

If this is not the case, check whether you assigned a value to the variable at some earlier time. For example, you may have assigned

```
x := 5;
```

earlier, then after awhile you typed:

```
expand((x-3)^2*(4*x+5));
```
$$100$$

You asked *Maple* to expand $(5-3)^2*(4*5+5) = 100!!$ To correct this, type:

```
x := 'x':
```

and reenter the **expand** command.

Question...   When I used **simplify** or **expand** on a simple polynomial, I got the wrong answer. What went wrong?

Answer...   Check to see that you remembered to put an asterisk * between terms being multiplied together. Parenthesized expressions don't get multiplied when they're written next to each other. Consider:

```
expand((x+1)(x+2)^2);
```
$$x(x+2)^2 + 2x(x+2) + 1$$

Maple interprets the expression **(x+1)(x+2)^2** as the function $(x+1)^2$ evaluated at $(x+2)$. This is not what you wanted to do.

Question...   I tried **simplify, expand, factor, combine,** and several other commands on an expression, and I can't get the type of expression I'm looking for to come out. What should I do?

Answer...   There are more advanced techniques for controlling the way *Maple* simplifies an expression. But the practical answer may be that the software just can't get you to where you want to be using algebra alone.

You must look for a different approach. For example, *Maple* can't simplify $\tanh^{-1}\big((e^x - e^{-x})/(e^x + e^{-x})\big) = x$. But by graphing $\tanh^{-1}\big((e^x - e^{-x})/(e^x + e^{-x})\big)$, you see that it looks like $y = x$. (Chapter 8 does graphing.) You can also compute that $\tanh^{-1}\big((e^x - e^{-x})/(e^x + e^{-x})\big)$ has derivative 1, so that's almost enough to establish the identity. (Chapter 11 will show you how to do derivatives.)

CHAPTER 5

Working with Equations

Equations and Their Solutions

The solve Command for an Equation

Maple's **solve** command will solve an equation for an "unknown" variable. You use it in the form:

> **solve(** *an equation* , *variable to solve for* **);**

For example,

```
solve( 2*x+5 = 9, x );
              2
```

Notice that equations in *Maple* are written using the equal sign "=".

You can check that $x = 2$ is the correct solution to the above equation, using substitution syntax:

```
subs(x=2, 2*x+5 = 9);
         9 = 9
```

You can even have *Maple* check that this substitution gives the correct answer by evaluating the result as a Boolean expression.

```
evalb(subs(x=2, 2*x+5 = 9));
              true
```

This means that after the substitution $x = 2$, the left-hand side of the equation equals the right-hand side.

Here are a few more examples involving **solve**:

Equation	To Solve It in Maple	Comment
Solve $x^2 - 3x + 1 = 0$ for x	`solve(x^2-3*x+1 = 0, x);` $$\frac{3}{2} + \frac{1}{2}\sqrt{5}, \ \frac{3}{2} - \frac{1}{2}\sqrt{5}$$ `evalf(")` #or **evalf(%);** for R5. 2.618033989, .381966011	Equations can have more than one solution. We can see a numerical answer with the **evalf** command.
Solve $x^3 + x^2 = -3x$ for x	`solve(x^3+x^2 = -3*x,x);` $$0, \ -\frac{1}{2} + \frac{1}{2}I\sqrt{11}, \ -\frac{1}{2} - \frac{1}{2}I\sqrt{11}$$	Here two of the solutions are complex. I stands for $\sqrt{-1}$.
Solve $y^2 - ay = 2a$ for y.	`solve(y^2 - a*y = 2*a, y);` $$\frac{1}{2}a + \frac{1}{2}\sqrt{a^2 + 8a}, \ \frac{1}{2}a - \frac{1}{2}\sqrt{a^2 + 8a}$$	If the equation involves other variables, *Maple* will treat them as constants.

| Solve $x + \sin x = \cos x$ for x | `solve(x+sin(x) = cos(x), x);`
(No output from *Maple*.) | No solution is returned when *Maple* cannot solve an equation. |

> **Note:** The **solve** command works very well for equations involving polynomials. However, it doesn't have much success with trigonometric, exponential, logarithmic, or hyperbolic functions.

The solve Command for a System of Equations

The **solve** command can also be used to solve a system of equations. For two equations in two unknown variables, you use this form of the command:

`solve({ equation1, equation2 }, { variable1, variable2 });`

■ **Example.** To solve the equations $3x + 8y = 5$ and $5x + 2y = 7$ in the variables x and y, use:

`solve({3*x+8*y = 5, 5*x+2*y = 7}, {x,y});`

$$\{x = \frac{23}{17}, \ y = \frac{2}{17}\}$$

■ **Example.** To solve the equations $3xy - y^2 = -4$ and $2x + y = 3$:

`solve({3*x*y - y^2 = -4, 2*x + y = 3}, {x,y});`

$$\{y = \text{RootOf}(5\_Z^2 - 9\_Z - 8), \ x = -\frac{1}{2}\text{RootOf}(5\_Z^2 - 9\_Z - 8) + \frac{3}{2}\}$$

Maple gives the answer in terms of the root of the polynomial $5z^2 - 9z - 8$. Since the polynomial is quadratic, we expect two sets of solutions. To get a list of solutions, we use the **allvalues** command:

`allvalues(");`   # or `allvalues(%);` for Release 5.

$$\{ \ x = \frac{21}{20} - \frac{1}{20}\sqrt{241}, \ y = \frac{9}{10} + \frac{1}{10}\sqrt{241} \ \},$$

$$\{ \ x = \frac{21}{20} + \frac{1}{20}\sqrt{241}, \ y = \frac{9}{10} - \frac{1}{10}\sqrt{241} \ \}$$

■ **Example.** If you try

`solve({x+y = 0, x+y = 1},{x,y});`

No solution is returned because there is no solution to this system of equations.

We can use a similar syntax to solve systems of equations involving three or more variables. For example:

`allvalues(solve({x+2*y-z=1, x-y+z^2=2, y+x=2*z},`
`{x,y,z}));`

$$\{z = -2 + 2\sqrt{2}, \ y = 3 - 2\sqrt{2}, \ x = -7 + 6\sqrt{2}\},$$
$$\{z = -2 - 2\sqrt{2}, \ y = 3 + 2\sqrt{2}, \ x = -7 - 6\sqrt{2}\}$$

Numerical Solutions for Equations

The fsolve Command for an Equation

The **fsolve** command uses efficient numerical techniques to approximate roots of polynomials and a few other simple functions. It has the same syntax as **solve**.

```
fsolve( an equation , variable to solve for );
```

```
fsolve(x^5-x^3-1=0,x);
```
> 1.236505703

The **fsolve** command gives a single answer. You can indicate the number of roots to look for with the **maxsols** option. Also we allow complex roots with the **complex** option. (Note that a fifth degree equation has at most five roots.)

```
fsolve(x^5-x^3-1=0,x, complex, maxsols=5);
```
> $-.9590477179 - .4283659563\,I,\quad -.9590477179 + .4283659563\,I,$
> $.3407948662 - .7854231030\,I,\quad .3407948662 + .7854231030\,I,\quad 1.236505703$

The fsolve Command for a System of Equations

The **fsolve** command can also do systems of equations, just like **solve**. You use the form:

```
fsolve({ equation1, equation2}, { variable1, variable2 });
```

For example,

```
fsolve({3*x*y - y^2 = 4, 2*x^2 + y = 9}, {x,y});
```
> $\{x = -2.946194209, y = -8.360120631\}$

```
fsolve({3*x*y - y^2 = 4, 2*x^2 + y = 9},
       {x,y},maxsols=7);
```
> $\{x = -2.946194209, y = -8.360120631\},\quad \{x = 2.031216860, y = .7483161321\},$
> $\{x = 1.614445936, y = 3.787128638\},\quad \{x = -2.199468588, y = -.6753241393\}$

If you tried **solve** with the system of equations above, the solution would involve the roots of a cubic equation. Using **allvalues** to convert that answer would produce several pages of output!

Guiding fsolve with a Range

fsolve with a Range in an Equation

Sometimes we are interested in a root other than the one that **fsolve** returns. We can guide **fsolve** by giving it a range for the solution. The form of the command becomes:

```
fsolve( equation , variable , range for the solution );
```

The **fsolve** command will then try to find a solution that is within the given range.

■ **Example.** We are interested in using **fsolve** to find solutions to the trig equation $\tan(x) = x$, If we know that there is a solution is each of the intervals $-\frac{\pi}{2} \le x \le \frac{\pi}{2}$,

$\frac{\pi}{2} \le x \le \frac{3\pi}{2}$, and $\frac{3\pi}{2} \le x \le \frac{5\pi}{2}$, we can specify appropriate ranges. We then find three solutions.

```
fsolve(tan(x) -x = 0, x, -Pi/2..Pi/2);
fsolve(tan(x) -x = 0, x, Pi/2..3*Pi/2);
fsolve(tan(x) -x = 0, x, 3*Pi/2..5*Pi/2);
```

$$0$$
$$4.493409458$$
$$7.725251837$$

Actually, by looking at the intersections of the graphs of $\tan(x)$ and x, we can see that there's a solution in the range $\frac{(2n-1)\pi}{2} \le x \le \frac{(2n+1)\pi}{2}$ for every n. You will learn how to draw this graph in Chapter 8, but for now, here's the picture:

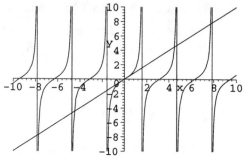

fsolve with a Range in a System of Equations

To solve a system of two equations for the two unknowns x and y, restricted to the intervals $x_0 \le x \le x_1$ and $y_0 \le y \le y_1$, you write:

```
fsolve({ equation1, equation2 }, {x,y},
    {x = x_0..x_1, y = y_0..y_1});
```

■ **Example.** The system $y^2 - x^3 = 5$ and $y = x - 3\cos x + 4$ has a solution inside each of the intervals $-2 < x < -1$ and $1 < y < 2$. We can pinpoint it with:

```
fsolve({y^2-x^3=5,y=x-3*cos(x)+4},{x,y},
    {x=-2..-1,y=1..2});
```

$$\{y = 1.769915245, x = -1.231437723\}$$

More Examples

Extracting Solutions from the Results of solve

Sometimes you will want to work with the answers you get from **solve** and **fsolve** without having to retype them. This example will show you the two steps that will enable you to do so.

Consider the equation $x^2 - 3x + 1 = 0$. It has two solutions:

```
solve(x^2-3*x+1 = 0, x);
```

$$\frac{3}{2} + \frac{1}{2}\sqrt{5}, \quad \frac{3}{2} - \frac{1}{2}\sqrt{5}$$

- Step 1. Make the output of the **solve** command the contents of a list and give the list a name. (We discuss lists in the next chapter.)

    ans := [solve(x^2-3*x+1 = 0, x)];

    $$ans := \left[\frac{3}{2} + \frac{1}{2}\sqrt{5}, \ \frac{3}{2} - \frac{1}{2}\sqrt{5} \right]$$

- Step 2. Identify each of the answers with **ans[1]** and **ans[2]**.

    ans[1];

    $$\frac{3}{2} + \frac{1}{2}\sqrt{5}$$

    ans[2];

    $$\frac{3}{2} - \frac{1}{2}\sqrt{5}$$

Let's check that **ans[2]** is really a solution to $x^2 - 3x + 1 = 0$:

    simplify(subs(x=ans[2],x^2 - 3*x + 1));

    $$0$$

How about:

    (x - ans[1]) * (x - ans[2]);

    $$\left(x - \frac{3}{2} - \frac{1}{2}\sqrt{5} \right) \left(x - \frac{3}{2} + \frac{1}{2}\sqrt{5} \right)$$

    expand(");   # or **expand(%);** for Release 5.

    $$x^2 - 3x + 1 = 0$$

Useful Tips

👁 👁 👁   **fsolve** works much more quickly using numeric methods than **solve** does algebraically. In fact, **solve** cannot solve many equations symbolically! Use **fsolve** whenever you can.

👁 👁   The following table can help you remember the difference between various types of brackets. But beware – they are not interchangeable!

Syntax Element	Purpose	Example
(*parentheses*)	(i) Grouping of terms in computation	(x^2+3)*(x–1)
	(ii) Used with commands	sin(x^2)
[*square brackets*]	List or vector	[x, y, z]
{ *curly braces* }	Grouping of expressions	{x–3 = y, 5*x+y = 2}
List [*square brackets*]	Specify a position in a list	soln[1]

Troubleshooting Q & A

Question... When I used **solve** or **fsolve**, I got the error message that said I had a "`syntax error`." What should I check?

Answer... You made a mistake in the input of the equation(s).

- Check that you entered the formula correctly. Common mistakes include misspelling names of the *Maple* built-in functions, and forgetting to type the multiplication symbol "`*`".

- If you are using more than one equation, make sure all the curly braces and commas are located in the right places.

Question... When I used **solve** or **fsolve**, I got the error message "`a constant is invalid as a variable`." What should I check?

Answer... Check that the variable or variables you are trying to solve for have no assigned values. For example, you may have earlier assigned:

```
x := 3 ;
```

Later, you try:

```
solve( x^2-1=0, x);
```

```
Error, (in solve) a constant is invalid as a variable, 3
```

You should have cleared the variable(s) before using **solve**.

```
x := 'x';
```

Question... I used **solve** but could not understand *Maple*'s output. It had the symbols `%1`, `%2`, and so on. What does this mean?

Answer... If the result from **solve** is too lengthy, *Maple* will use the symbols `%1`, `%2`, etc., to represent expansions of subexpressions that occur several times.

Question... I used **solve** and *Maple* gave me the output 'RootOf.' What does this mean?

Answer... For example, if you try:

```
sol := [solve( sqrt(x) = (x-3)^2 , x )];
```

$$sol := [\ \text{RootOf}(-\_Z + \_Z^4 - 6\_Z^2 + 9,\ 2.110124849)^2,$$
$$\text{RootOf}(-\_Z + \_Z^4 - 6\_Z^2 + 9,\ 1.354977808)^2\ ]$$

Maple tells you that the solutions to the equation $\sqrt{x} = (x-3)^2$ are the squares of the roots of the polynomial $-z + z^4 - 6z^2 + 9$ near $z = 2.11$ and 1.35. This is not helpful. However, you can see the numerical value of these answers with:

```
evalf({sol[1], sol[2]} );
```

$$\{4.452626878, 1.835964860\}$$

If the equation involves polynomial only, you can also use the **allvalues** command to convert the answer to a friendlier form:

allvalues(");   # or **allvalues(%)** for Release 5.

Question... *Maple* did not give any output from **solve** or **fsolve**. Why not?

Answer... This usually means one of three things.

- *Maple* does not know how to solve your equation(s) with the **solve** command.

- If you used **fsolve** with a specified interval, *Maple* was unable to locate a root in that interval. Make sure that the interval(s) you gave contains a root.

- There is no solution to your equation or system of equations. A single equation might reduce to an absurdity (e.g., $x = x + 1$ reduces to $0 = 1$). A system of equations may be inconsistent (e.g., $x + y = 1$ and $x + y = 2$).

Question... When I used **fsolve**, I got the error message that I "...should use exactly all the indeterminates." What happened?

Answer... *Maple* cannot use a numerical method to approximate the solution, because your equation involves a constant that is not defined. For example, if **a** is not defined, *Maple* can do:

solve(x^2 = a, x);

But you will get an error message with:

fsolve(x^2 = a, x);

Question... **fsolve** failed to give me a solution to an equation in an interval where I know there is a solution. What happened?

Answer... **fsolve** uses a version of Newton's method to search for an answer. The procedure will not always find an answer. It is particularly vulnerable to functions whose graphs have narrow spikes on otherwise well-behaved regions.

For example, away from the origin, $\tan(x) = x^3$ has this feature, and it is hard for **fsolve** to locate a root without a very precise range.

CHAPTER 6
Sets, Lists and Sequences

Lists and Sets

What Is a List?

A **list** in *Maple* is an expression in which elements are separated by commas and enclosed in [*square brackets*]. For example, each of the following is a list.

```
[2, 5, 7, 10, -3, -25];        # Each element is a number.

[`good`, `bad`, `ugly`];       # Each element is a name or string.

[[1,3], [2,1], [5,6]];         # Each element is a list of numbers.
```

What Is a Set?

A **set** in *Maple* is an expression in which elements are separated by commas and enclosed in { *curly braces* }. For example, each of the following is a set.

```
{2, 5, 7, 10, -3, -25};        # Each element is a number.

{`good`, `bad`, `ugly`};       # Each element is a name or string.

{[1,3], [2,1], [5,6]};         # Each element is a list of numbers.
```

Lists and sets are important structures in *Maple*. Many of *Maple's* inputs and outputs are expressed using lists or sets. For example, *Maple* gives us two sets for the output when we solve a system of equations:

```
solve({x+y=2, x^2+y = 2}, {x,y});
```
$$\{x = 1, y = 1\}, \ \{x = 0, y = 2\}$$

We can then collect the solutions to form a list:

```
ans :=   [ solve({x+y=2, x^2+y = 2}, {x,y}) ];
```
$$ans := [\{x = 1, y = 1\}, \{y = 2, x = 0\}]$$

What Is the Difference?

A **list** is an ordered set. We can have repetition in the elements. In a **set** on the other hand, the ordering and redundancy of elements does not matter.

```
{5,4,3,2,1};      # The arrangement of the elements in a set does not matter.
```
$$\{1, 2, 3, 4, 5\}$$

Also, duplicate elements are automatically removed in a set but not in a list:

```
{1,2,2,3,3,3,4,4,4,4,5,5,5,5,5};
```
$$\{1, 2, 3, 4, 5\}$$

```
[1,2,2,3,3,3,4,4,4,4,5,5,5,5,5];
```
$$[1, 2, 2, 3, 3, 3, 4, 4, 4, 4, 5, 5, 5, 5, 5]$$

Sequences

The seq Command	The **seq** command can be used to generate a sequence of elements that can be defined by a mathematical formula. For example, to make a sequence of the form n^2, for each integer $1 \le n \le 16$, we will type:

> **`seq(n^2, n=1..16);`**
>
> $1, 4, 9, 16, 25, 36, 49, 64, 81, 100, 121, 144, 169, 196, 225, 256$

In general, the **seq** command is used in the form:

> **`seq( ` *expression* `, n = ` n_0 `.. ` n_1 `);`**

Here are a few examples:

Maple *Command*	*Remark*
`seq(n^2, n=-2..8);` $4, 1, 0, 1, 4, 9, 16, 25, 36, 49, 64$	The variable n runs from -2 to 8.
`seq([cos(n),n/(n+1)], n = 1..3);` $\left[\cos(1), \dfrac{1}{2}\right], \left[\cos(2), \dfrac{2}{3}\right], \left[\cos(3), \dfrac{3}{4}\right]$	The expression in the **seq** command can be a list itself. In this case, we form a table of coordinate pairs.
`seq(x^n,n=0..7);` $1, x, x^2, x^3, x^4, x^5, x^6, x^7$	Each element can be a symbolic expression.

By combining the **seq** command with [*square brackets*] or { *curly braces* }, we can create a list or set. For example,

> **`[seq(n, n=1..22)];`**
>
> $[1, 2, 3, 4, 5, 6, 7, 8, 9, 10, 11, 12, 13, 14, 15, 16, 17, 18, 19, 20, 21, 22]$

> **`{seq(cos(n), n =1..5)};`**
>
> $\{\cos(5), \cos(4), \cos(3), \cos(1), \cos(2)\}$

More Examples

An Experiment in Factoring	Let's look at the factorization of $x^n + 1$. With the help of the **seq** command, we can see the factorization for several values of n very quickly, say for $n = 2, 3, \ldots, 10$.

> **`seq(factor(x^n+1), n=2..10);`**
>
> $x^2 + 1, \ (x+1)\,(x^2 - x + 1), \ x^4 + 1, \ (x+1)\,(x^4 - x^3 + x^2 - x + 1),$
>
> $(x^2 + 1)\,(x^4 - x^2 + 1), \ (x+1)\,(1 - x + x^2 - x^3 + x^4 - x^5 + x^6), \ x^8 + 1,$
>
> $(x+1)\,(x^2 - x + 1)\,(x^6 - x^3 + 1), \ (x^2 + 1)\,(x^8 - x^6 + x^4 - x^2 + 1)$

Do you notice that all the coefficients in all the factors above are either 1 or -1? Try repeating the command with **n = 2..100**. You will see the same thing!

You may probably conclude that every coefficient in every factor of $x^n + 1$ will be ±1, no matter what n is. But before you celebrate your latest discovery, check the factorization of $x^{105} + 1$. Surprise!

Drawing a Nonagon

■ **Example.** A nonagon is a regular polygon of nine sides (i.e., all nine sides have the same length and all nine angles are equal). It can be formed by joining together the points with coordinates $(\cos(2\pi n / 9), \sin(2\pi n / 9))$, where $n = 1, 2, \ldots, 9$.

We can create a list of these points using the **seq** command:

```
coordlist:=
        [seq([cos(n*2*Pi/9),sin(n*2*Pi/9)],n=0..9)];
```

$coordlist := [[1, 0], [\cos(\tfrac{2}{9}\pi), \sin(\tfrac{2}{9}\pi)], [\cos(\tfrac{4}{9}\pi), \sin(\tfrac{4}{9}\pi)], [\tfrac{-1}{2}, \tfrac{1}{2}\sqrt{3}],$
$[-\cos(\tfrac{1}{9}\pi), \sin(\tfrac{1}{9}\pi)], [-\cos(\tfrac{1}{9}\pi), -\sin(\tfrac{1}{9}\pi)], [\tfrac{-1}{2}, \tfrac{-1}{2}\sqrt{3}],$
$[\cos(\tfrac{4}{9}\pi), -\sin(\tfrac{4}{9}\pi)], [\cos(\tfrac{2}{9}\pi), -\sin(\tfrac{2}{9}\pi)], [1, 0]]$

If we join the points one by one, we will create the nonagon. This can be done using the **plots[pointplot]** command. (**pointplot** will be discussed in Chapter 21.)

```
plots[pointplot]( coordlist, style=line,
                          scaling=constrained);
```

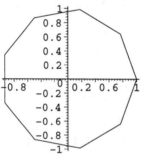

(The option **scaling = constrained** is used to make sure that both x- and y-axes have the same unit scale. We will explain this in detail in Chapter 8.)

Troubleshooting Q & A

Question... When I used the **seq** command, *Maple* gave me no output. What does that mean?

Answer... This means that the sequence you asked for is empty, because the index range you specified does not make sense. Make sure that $n_0 \le n_1$ in the command

```
seq( expression, n=n0..n1);
```

CHAPTER 7

Getting Help and Loading Libraries

Text-based Help

Help for Specific *Maple* Commands

No matter which version of *Maple* you use, you can always get some help on how to use a command by typing:

??*command*

For example, if you forget how to use the **fsolve** command, you can type:

??fsolve
```
CALLING SEQUENCE:
    fsolve( <eqns>, <vars>, <options> );
PARAMETERS:
    <eqns> -  an equation or set of equations
     <vars> - (optional) an unknown or set of unknowns
    <options> - (optional) parameters controlling solutions
```

If you type **?** with a command name, you get more information. In a graphically based system, a help window will open, giving you details of the command, including options and examples. For example:

?fsolve

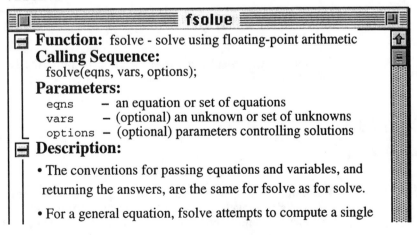

Function: fsolve - solve using floating-point arithmetic
Calling Sequence:
    fsolve(eqns, vars, options);
Parameters:
    eqns   – an equation or set of equations
    vars   – (optional) an unknown or set of unknowns
    options – (optional) parameters controlling solutions
Description:
  • The conventions for passing equations and variables, and returning the answers, are the same for fsolve as for solve.

  • For a general equation, fsolve attempts to compute a single

Looking Up a Command

Sometimes, you won't remember the exact name of a command that you want to use. Say, you remember that it started with **plo**, but you just can't recall its full name. You can ask *Maple*:

```
?plo
Try one of the following topics:
     {plotsetup, PLOT3D, plot, plot3d, plots, PLOT}
```

Browser-based Help

For a graphically based system, you can open up the interactive Help Window from the Help menu. To look up information, you check the list of categories and select the appropriate one. (The following picture shows how the contents of the Help window looks like for Release 4 on a Mac.)

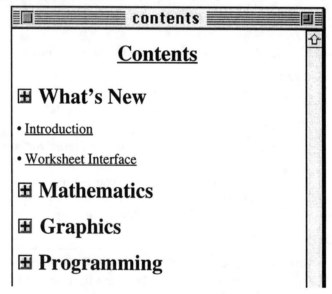

The information displayed from the Help menu is the same as what you get from using ?*command*.

Packages

Loading a Library

Maple has a built-in vocabulary of several hundred commands. But still these are not enough for everyday usage. Additional commands are available in the "Library packages." These include commands for Algebra, Calculus, Linear Algebra, Number Theory, Statistics, and so on.

Before using those commands, you have to load their corresponding library using the **with** command. For example, the **animate** command is defined in the **plots** library. We will load it with:

`with(plots);`

[*animate, animate3d, conformal, contourplot, cylinderplot, densityplot, display, display3d, fieldplot, fieldplot3d, gradplot, gradplot3d, implicitplot, implicitplot3d, loglogplot, logplot, matrixplot, odeplot, pointplot, polarplot, polygonplot, polygonplot3d, polyhedraplot, replot, setoptions, setoptions3d, spacecurve, sparsematrixplot, sphereplot, surfdata, textplot, textplot3d, tubeplot*]

It lists all the commands defined in the **plots** library that have been loaded.

Now we can use the **animate** command:

`animate(sin(2*x*t), x=-2..2, t=-2..2);`

Now, click on the picture, then click the play button ▶ and the show starts! (Chapter 23 discusses animation in detail.)

Use a Command Without Loading a Library

You can also load and use a single command from a library without loading the whole library. For example, if you only want to use the **polygonplot** command defined in the **plots** library, you can type:

`plots[polygonplot]` (*input values for the command*)`;`

Maple will then know where to find the definition of that command. However, since you have not loaded the **polygonplot** command, the next time when you use it, you have to type **plots[polygonplot]** again.

As another example, if you want to use the **mean** command which is defined in the sublibrary **describe** of the **stats** library, you type:

`stats[describe, mean]` (*input values for the command*)`;`

Useful Tips

💡 💡 Most of the help pages contain examples of how commands can be used. Pick the one or two closest to what you are trying to do, copy them to your worksheet, and execute them to see how the commands work.

💡 When you load a library, finish the command with a colon instead of a semicolon,

`with(` *library*)`:`

Then *Maple* will not display the whole list of the commands in that library, which can be very lengthy.

CHAPTER 8

Making 2-D Pictures

Drawing the Graph of a Function

The plot Command

A common operation in mathematics is to plot the graph of a function, such as that of the square function, over a given interval. *Maple* does this with its **plot** command To see the graph of $f(x) = x^2$ over the interval $-3 \le x \le 2$, type:

plot(x^2, x=-3..2);

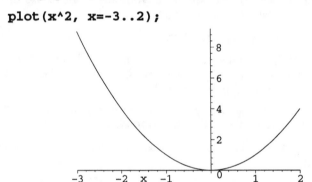

In general, to plot a function of x over an interval $a \le x \le b$, you type:

plot(*function*, x = *a..b*) ;

For example, this plots the graph of $y = x^3 - 3x^2 + 5x - 10$ for x in the interval $-2 \le x \le 3$.

plot(x^3 -3*x^2 + 5*x -10,   x=-2..3);

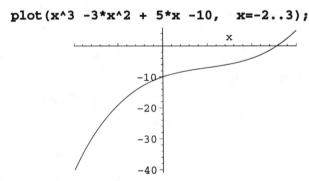

Notice that if you click the mouse with the cursor/arrow at a particular point, the box at the bottom (or at the top, depending on your machine and version of *Maple*) of the graph will give the coordinates of that point.

Plot Options Tool Bar

In *Maple V* release 4, when you click once on the center of a plot, the plot options tool bar will show up near the top of the worksheet. By clicking various buttons in the options bar, you can select different plot styles, axes styles and scales for the picture.

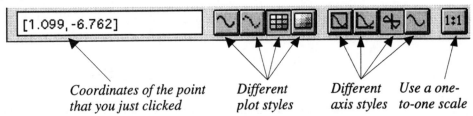

Coordinates of the point that you just clicked   *Different plot styles*   *Different axis styles*   *Use a one-to-one scale*

Also, you can redraw the graphic by choosing various options under the **Style** menu, **Axes** menu and **Projection** menu on the menu bar.

Plotting Multiple Functions

The **plot** command allows you to graph several functions or expressions simultaneously, all on the same set of axes, over a common interval $a \le x \le b$. You use the following format.

```
plot ( { function1, function2, etc. }, x = a..b);
```

For example, to see the graphs:

$$y = 3x^4 - 5x, \quad y = 10\sin(x) - 10, \quad y = 5\cos(x) + 3e^x$$

on the interval $-2 \le x \le 2$, you type:

```
plot({3*x^4-5*x, 10*sin(x)-10, 5*cos(x) + 3*exp(x)},
     x = -2..2);
```

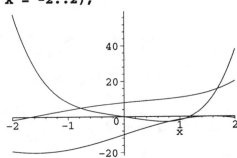

Notice that we use { *curly braces* } to group the functions together. The only drawback to this command is that all the functions must have the same domain (in this case **x=-2..2**). Later in this chapter, we will show you another method to combine different pictures.

Scaling Option

Sometimes picture output is not scaled correctly. Consider the following graphs:

$$y = \sqrt{1 - x^2} \quad \text{and} \quad y = -\sqrt{1 - x^2}$$

```
plot( {sqrt(1-x^2), -sqrt(1-x^2)}, x=-1..1);
```

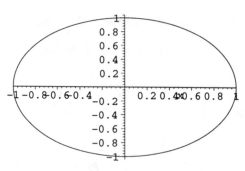

Notice that the formulas $y = \sqrt{1-x^2}$ and $y = -\sqrt{1-x^2}$ come from the equation $x^2 + y^2 = 1$ which defines the unit circle. But what you see from the graph appears to be an ellipse instead. This is because the computer uses different unit lengths for the x- and y-axes. To correct this, you can either resize the picture (by dragging its corner with the mouse) or type:

```
plot({sqrt(1-x^2), -sqrt(1-x^2)}, x=-1..1,
     scaling = constrained);
```

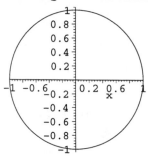

Now both axes will have the same scale. (In release 4 or higher, you can also click on the "**1:1**" button in the plot option menu bar to obtain the same scaling.)

> **Note**: You should use **scaling = constrained** whenever you want to see angles or circles properly. For example, the graph of $f(x) = x$ will not look like it makes a $45°$ degree angle with the x-axis unless you specify **scaling = constrained**. (See the pictures below.)

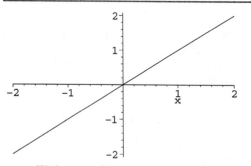

Without **scaling = constrained**

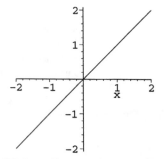

With **scaling = constrained**

Restricting the y-axis   When a graph contains a vertical asymptote or has a very large range of y-values, some interesting features of the graph may not be visible. This is because *Maple* tries to show you the whole graph. The solution is to restrict the portion of the *y*-axis that is shown.

For example, the graph of the tangent function has vertical asymptotes. But consider the difference between *Maple*'s default picture and one where we restrict the graph to show only *y*-values with $-10 \leq y \leq 10$.

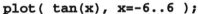

`plot( tan(x), x=-6..6 );`

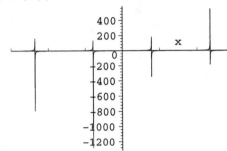

`plot( tan(x), x=-6..6, y=-10..10 );`

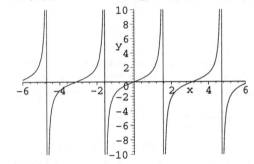

Working with Asymptotes and Discontinuities

Maple draws asymptotes in the picture above because it actually tries to connect the branches of the graph together, from top to bottom. You can turn this behavior off by setting **discont = true** in the **plot** command.

`plot( tan(x), x=-6..6 , y=-10..10, discont=true);`

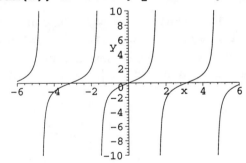

Labeling Pictures

You can add a label to a picture and change the labels on the axes by using the **title** and **labels** options in the **plot** command.

```
plot( 15+cos(x), x=0..4*Pi ,
   labels = [ `day`, `price` ],
   title = `Daily price of Stock ABC` );
```

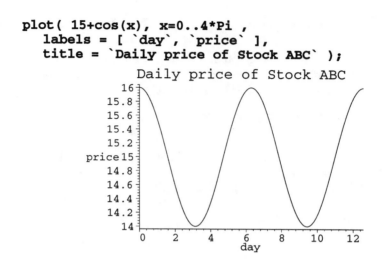

More Advanced Drawings

Combining Existing Plots with the display Command

You can combine different graphics into a single picture by using the **display** command defined in the **plots** library. First, name each of the graphics. Then, use the **display** command to combine these named graphics into a single output.

```
graph1 := plot( cos(x),x=-3..1):      # Note we use : at the end
                                      #  of the command.
graph2 := plot(sin(x),x=0..3):
graph3 := plot([[-2,0.2],[3,0.5]], x=-2..4):
```

We conclude each command with a colon instead of a semicolon. The colon serves the same purpose as a semicolon, except that the colon indicates that *Maple* should perform the desired calculation internally without displaying the result. In this case, if we had used a semicolon, we could have still stored the plots, but then *Maple* would have displayed a long list of (unimportant) calculations from the **plot** commands.

```
with(plots):          # We have to open the plots library to use display.
display({graph1, graph2, graph3});
```

Notice that we end this command with a semicolon this time, because we want *Maple* to show us the picture. If you use a colon here, you will not see the picture.

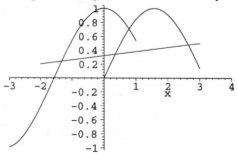

Colors

You can change the color of a picture by specifying the **color** option inside the **plot** command. You have a wide variety of colors to choose from, such as black, blue, navy, coral, cyan, brown, gold, green, gray, maroon, orange, pink, plum, red, sienna, tan, turquoise, violet, wheat, white, and so on.

Say, let us draw a picture in orange:

```
plot(x^3+2*x, x=-3..2, color = orange);
```

Sorry, we cannot show you the picture here, because this text is printed in black and white!

Thickness

You can change the thickness of a graph by setting the **thickness** option to have the value 1 for a thin pen, 2 for a medium pen, or 3 for a thick pen. (The default is to use a thin pen.)

```
plot(x^3+2*x, x=-3..2, thickness = 3);
```

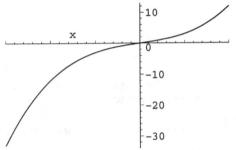

More Examples

Zooming In

■ **Example.** Consider $f(x) = \sqrt{1 + 10x^4 - 20x^5 + 25x^6}$ and $g(x) = x^2 + 5\sin(x)$:

```
f := x -> sqrt( 1 + 10*x^4 - 20*x^5 + 25*x^6 );
g := x -> x^2 + 5*sin(x);

plot({f(x),g(x)}, x=-3..3);
```

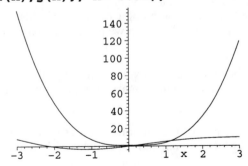

To get a better estimate of the intersection near 1.1, we can zoom in on the graph by successively shrinking down the x-interval in the **plot** command.

```
plot({f(x),g(x)}, x=1 .. 1.2);
plot({f(x),g(x)}, x=1.13 .. 1.15);
plot({f(x),g(x)}, x=1.135 .. 1.145);
```

"Optical Illusion"

■ **Example.** Suppose we plot the function $f(x) = 64x^4 - 16x^3 + x^2$:

```
f := x -> 64*x^4-16*x^3+x^2;
```

```
plot(f(x), x=0..10);
```

You can easily get the impression that the graph is always increasing. However, notice that the vertical range is very large, between 0 and 600,000, so the picture is not sharp enough to show any small dips. In particular, the graph looks like a straight line for x between 0 and 2; this certainly is not the case.

```
plot(f(x), x=0..2);        # Still looks OK.
plot(f(x), x=0..1);        # Still looks OK.
plot(f(x), x=0..0.5);      # Still looks OK.
plot(f(x), x=0..0.2);      # Surprise!!
```

Useful Tips

☿ ☿ ☿   *Maple* uses a variety of colors to draw 2-D graphs. These may look pleasing on screen, but may not look very good when printed on a gray scale printer. We suggest that unless you have a color printer available, you add the option **color = black** to your plots (as we did when we printed the pages of this text).

☿ ☿   If you have a **plot** that involves various complicated functions, always try it first with one function at a time to make sure the formula works.

☿   There are many other options in **plot** that we have not discussed. You can find them using **?plot[options]** and experiment with them.

Troubleshooting Q & A

Question... I got an empty picture from **plot** with an error message "Warning in iris-plot: empty plot." What went wrong?

Answer... This indicates that *Maple* cannot evaluate your input function numerically. Check whether you made a typo in the input. The common mistakes are:

- You typed the name of a built-in function incorrectly.
- You used the wrong variable.
- You specified an interval in which the input function is not undefined.

Check whether your function really gives numbers! Do you get numbers when you enter **f(–1), f(0), f(1)** and so on?

Question... When I combined various pictures using the **display** command, I got a long list of numbers but no picture. Where is my mistake?

Answer... Make sure that you load the **plots** library before using the **display** command. Type:

```
with(plots);
```

and reenter your **display** command.

Also check that you have entered the format correctly. Note that you have to include all the names of the graphics inside a { *curly braces* } in the **display** command.

Question... When I combined various pictures using the **display** command, I got an error message. Where is my mistake?

Answer... Check if you typed the names of the pictures correctly in your **display** command. If there is no misspelling there, then you should **display** each picture one at a time in order to find out which one is causing the problem. Then recheck their definitions. (A common mistake is to use = instead of := in defining the pictures.)

CHAPTER 9

Plotting Parametric Curves and Line Segments

Parametric Curves

Plotting Parametric Curves

In the previous chapter you saw how to use **plot** to draw curves that are graphs of functions. But not all curves are the graphs of functions.

A two-dimensional (2-D) parametric curve is written in the form $(x(t), y(t))$. The **plot** command can also be used to draw the curve $(x(t), y(t))$, when t varies from a to b. The command has the form:

```
plot([ x(t), y(t), t = a..b ]);
```

For example, to see the curve $(t^2 - 1, t + t^2)$ for $-2 \leq t \leq 2$, we use:

```
plot([t^2-1, t+t^2, t=-2..2]);
```

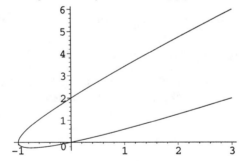

Plot Options

The options that we discussed in the last chapter (e.g., **scaling = constrained** and **color**) will also work for this particular format of **plot** command.

■ **Example**. To see the unit circle $(\cos t, \sin t)$ for $0 \leq t \leq 2\pi$, you use:

```
plot([cos(t),sin(t), t=0..2*Pi]);
```

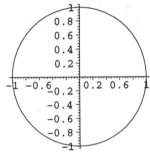

Maple will give you a picture that does not look like a circle because the two axes are not measured in the same unit length. To correct this, we add the option **scaling = constrained**. We will also draw it with a blue color:

```
plot([cos(t),sin(t), t=0..2*Pi],
     scaling = constrained, color = blue);
```

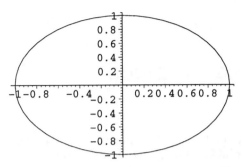

Plotting Line Segments in 2-D Plots

Line Segment   The **plot** command can also be used to plot a line segment joining the points (a, b) and (c, d). You type:

```
plot( [ [a, b], [c, d] ], x = x₀ .. x₁);
```

For example, to see the line segment joining $(0, 1)$ to $(-1, 3)$:

```
plot([[0,1],[-1,3]], x=-2..2) ;
```

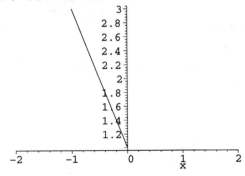

The previous command will only work for Release 3 or higher. In Releases 1 and 2, you have to use the following form of the command:

$$\texttt{plot(} \; [a, \; b, \; c, \; d] \; \texttt{, } \; \texttt{x} \; \texttt{=} \; x_0 \texttt{..} x_1\texttt{);} \qquad \text{\# For R1 or R2.}$$

Plotting Multiple Graphics

Multiple Curves

You can plot several parametric curves, graphs, and lines all in one picture, using a single **plot** command. To do this, you put all the expressions of the curves inside { *curly braces* }.

■ **Example**. To draw the curves (t^2, t) for $-2 \le t \le 2$, and (t, t^2) for $-1 \le t \le 1$, the line segment from (1, 0) to (2, 3), and the graphs $y = 2x$, $y = x^2$ for $-2 \le x \le 3$, we can use:

```
plot({[t^2,t, t=-2..2] ,[t,t^2, t=-1..1],
      [[1,0],[2,3]], 2*x, x^2 }, x=-2..3);
```

Multiple Pictures Together

You can also combine several 2-D pictures with the **display** command inside the **plots** library. The pictures you combine do not need to have the same interval for x. Also, we can assign different options to each individual picture.

■ **Example**. To create the "open skull" picture that you see below, we will combine several pictures. The face is made from a parabola and two straight lines; the eyes from two circles; and the mouth from a straight line.

```
face := plot({x^2, 4, 4.5+x/4},x=-2..2, color = red):
eyes := plot({[1+cos(t)/4, 3+sin(t)/4, t=0..2*Pi],
              [-1+cos(t)/2, 3+sin(t)/2, t=0..2*Pi]},
             color = brown):
mouth := plot([[-0.5,1],[0.5,1]], color = blue):

with(plots):
display({face,eyes,mouth}, scaling=constrained);
```

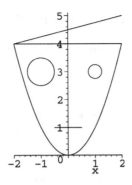

Useful Tips

☿ ☿ ☿   You may be overwhelmed by all the different plotting formats that we have discussed. The following table can help you to remember them:

Expression of the Form	*What It Plots*
plot({ $f_1(x)$, $f_2(x)$}, x = a..b);	The graphs of $f_1(x)$, $f_2(x)$
plot([$x(t)$, $y(t)$, t = a..b]);	The parametric curve $(x(t), y(t))$
plot([[a, b], [c, d]]);	The line segment (a, b) to (c, d)

Always remember that [*square brackets*] are generally related to coordinates.

☿ ☿   If you need to draw a complicated picture that consists of various graphs or curves, always draw each picture individually and combine them with the **display** command. You may be able to use a single **plot** command to draw these graphics, but if you get an error message, it will be difficult to locate a mistake.

Troubleshooting Q & A

Question...   I tried to draw a parametric curve with **plot** but got an error message "Warning in iris-plot: empty plot" or "Plotting error: empty plot." What went wrong?

Answer...   This indicates that *Maple* cannot evaluate your input function numerically. Check whether you made a typo in the input. Common mistakes are:

- You typed the wrong name of a built-in function.
- You used the wrong variable.
- You specified an interval in which the input function is not undefined.

Question...   I tried to draw a parametric curve with **plot** but got an error message "Error, (in plot) invalid arguments." What should I check?

Answer...   This error message suggests that you should check the format of the **plot** command. Make sure that you used [*square brackets*] and (*parentheses*) correctly. For example, a common mistake is to input the expression **(t, t^2)** or **(t, t^2, t = –2..2)** instead of **[t, t^2, t = –2..2]** for the curve (t, t^2).

Question...   I tried to draw a parametric curve with **plot** but got an error message "`Error, (in plot) parameter range must evaluate to a numeric.`" What should I check?

Answer...   This error message suggests that you made a mistake in specifying the range for the parameter. For example, check to see if you used **pi** instead of **Pi**.

Question...   I tried to draw one parametric curve with **plot**, but instead *Maple* gave me a picture of two curves. What went wrong?

Answer...   When entering a parametric curve, it is common to forget to include the interval *inside* the [*square bracket*]. For example, instead of typing **plot([t, t^2, t = –2..2])**; you may make the mistake of typing:

```
plot([t, t^2], t=-2..2);
```

In *Maple V* Release 4 or higher, this command will be interpreted as *similar* to:

```
plot({t, t^2}, t=-2..2);
```

Then *Maple* will draw the *graphs* $y = t$ and $y = t^2$!! This is not what you want.

CHAPTER 10
Polarplot and Implicitplot

Plotting in Polar Coordinates

The polarplot Command

If a curve is expressed in polar coordinates, then we can draw it with the **polarplot** command. This command is defined in the **plots** library. (See the discussion of libraries in Chapter 6.)

If the curve is given by $r = f(\theta)$, for $\theta_1 \le \theta \le \theta_2$, we can draw it with the following.

```
with(plots):
polarplot( f(θ), theta = θ₁ .. θ₂ );
```

Notice that this form is very similar to that of the **plot** command you already know. For example, to plot the three-leaf rose $r = 2\cos(3\theta)$, use:

```
with(plots):
polarplot( 2*cos(3*theta) , theta = 0..2*Pi);
```

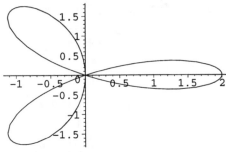

Multiple Plotting

Just as in the **plot** command, you can plot several curves by entering them inside { *curly braces* }.

For example, you can plot the spirals $r = \frac{\theta}{2\pi}$, $r = \left(\frac{\theta}{2\pi}\right)^2$ and the circle $r = 1$ all in one picture with:

```
polarplot({ theta/(2*Pi), (theta/(2*Pi))^2, 1 },
          theta = 0..2*Pi);
```

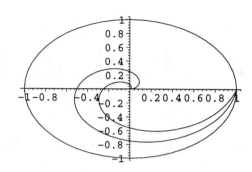

Plotting Graphs of Equations

The implicitplot Command

When a curve is given by an equation in variables x and y, you can draw the curve with the **implicitplot** command (defined in the **plots** library).

You use the command in the form:

```
with(plots):
implicitplot( an equation in x and y , x = a..b, y = c..d );
```

For example, to see the unit circle $x^2 + y^2 = 1$ for $-1 \le x \le 1$, $-1 \le y \le 1$ use:

```
with(plots):
implicitplot( x^2 + y^2 = 1, x=-1..1, y=-1..1);
```

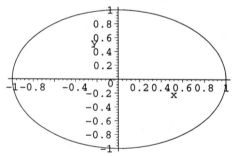

Multiple Curves and Styles

You can also use { *curly braces* } or the **display** command to plot multiple implicit curves. This is similar to the way you do it in the **plot** and **polarplot** commands.

■ **Example.** We want to see the curves $x^2 + 3xy + y^3 = 25$, $x^2 + 3xy + y^3 = 10$ and $x^2 + 3xy + y^3 = 0$ in the intervals $-10 \le x \le 10$ and $-6 \le y \le 4$. Let's draw each curve with a different color:

```
f := (x,y) -> x^2+3*x*y+y^3;
pict1 := implicitplot( f(x,y) = 25, x=-10..10,
          y=-6..4, color = red):
pict2 := implicitplot( f(x,y) = 10, x=-10..10,
          y=-6..4, color = yellow):
pict3 := implicitplot( f(x,y) = 0, x=-10..10,
          y=-6..4, color = green):
display({pict1,pict2,pict3});
```

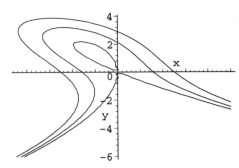

This picture looks like the Chinese character for "Wind."

Useful Tips

💡 💡 You may be overwhelmed by all the different plotting commands that we have discussed. Don't worry, **plot** is the most commonly used command, while **polarplot** and **implicitplot** are for special situations:

To Draw	Use
Graph of a function $f(x)$	`plot`
Parametric curve $(x(t), y(t))$	`plot`
Curve in polar coordinates	`polarplot`
Curve given by an equation	`implicitplot`

Troubleshooting Q & A

Question... I got an error message "Plotting error, empty plot," when I used **polarplot**. What happened?

Answer... Check whether you made a typo in the input. Common mistakes are:

- You mistyped the name of a built-in function/variable.

- Your input function contains a variable other than **theta** (a common mistake is to include **r**, the radius variable, in the input).

- You did not match up parameters correctly (e.g., you used **t** in the function but used **theta** when you specified the interval).

Question... I got an error message from **implicitplot**. What should I look for?

Answer... There are two major problem areas in using **implicitplot**.

- Make sure that the equation you entered is really an equation with an equal sign "=". Check that the equation is entered correctly. Did you remember to type * for multiplication?

- Make sure that your equation has two variables that do not have values.

Executing **x := 'x'; y := 'y';** before using **x** and **y** in your equation for **implicitplot** is highly recommended.

CHAPTER 11
Limits and Derivatives

Limits

The limit Command

If f is a function of a single variable, *Maple* evaluates $\lim\limits_{x \to a} f(x)$ with the following syntax:

limit(*function*, **x** = *a*)**;**

For example, *Maple* agrees that $\lim\limits_{x \to 0} \dfrac{\sin x}{x} = 1$

limit(sin(x)/x, x = 0);

$$1$$

The following table shows some sample limit computations, demonstrating that *Maple* can compute most limits, even those that involve infinite limits or limits at infinity:

Limit Calculation	*In* Maple
$\lim\limits_{x \to 0} \dfrac{e^x - 1 - x}{x^2} = \dfrac{1}{2}$	**limit((exp(x)-1-x)/x^2, x = 0);** $$\dfrac{1}{2}$$
$\lim\limits_{x \to 3} \dfrac{1 - x}{(x - 3)^2} = -\infty$	**limit((1-x)/(x-3)^2, x = 3);** $$-\infty$$
$\lim\limits_{x \to +\infty} \dfrac{x}{\sqrt{x^2 + 1}} = +1$ $\lim\limits_{x \to -\infty} \dfrac{x}{\sqrt{x^2 + 1}} = -1$	**f := x -> x/sqrt(x^2+1);** **limit(f(x),x = infinity);** $$1$$ **limit(f(x),x = -infinity);** $$-1$$
$\lim\limits_{x \to 0} \dfrac{\lvert x \rvert}{x}$ does not exist	**limit(abs(x)/x, x=0);** *undefined*

One-sided Limits

To calculate a left-hand limit $\lim\limits_{x \to a^-} f(x)$, you have to add the **left** option in the **limit** command. To find $\lim\limits_{x \to 0^-} \dfrac{\sqrt{2x^2}}{x}$, you use:

limit(sqrt(2*x^2)/x, x = 0, left);

$$-\sqrt{2}$$

Similarly, to find the right-hand limit $\displaystyle\lim_{x\to 0^+}\frac{\sqrt{2x^2}}{x}$, you use:

```
limit( sqrt(2*x^2)/x, x = 0, right );
```
$$\sqrt{2}$$

Limits and Graphs

■ **Example.** We compute:

```
limit(sin(sin(2*x)^2)/x^2, x = 0);
```
$$4$$

We can check both numerically and graphically to see if this answer is correct.

- (*Numerically*) Calculate the values of the function at various test values of x that are close to zero. Check if they approach 4:

```
f := x -> sin(sin(2*x)^2)/x^2;
```

```
[f(0.12), f(-0.1), f(0.0123), f(-0.0125),
   f(0.001234), f(-0.0011)];
```
$$[3.921699930,\ 3.945925593,\ 3.999192941,$$
$$3.999166474,\ 3.999991880,\ 3.999993545]$$

- (*Graphically*) Check from the graph if the height of the function approaches 4, as x approaches zero.

```
plot( sin(sin(2*x)^2)/x^2, x = -0.5..0.5);
```

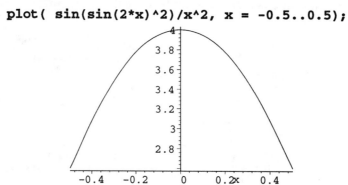

Differentiation

Differentiation Using the diff command

Maple uses **diff** for computing derivatives. You use it in the form:

```
diff( function, variable );
```

For example,

```
diff( x^4, x );
```
$$4x^3$$

To calculate a second derivative, you have to use **diff** twice:

```
diff( diff( x^4, x ), x);
```

$$12\,x^2$$

Or you can use either of these short-hand formats:

```
diff( x^4, x, x);
```

$$12\,x^2$$

```
diff( x^4, x$2);
```

$$12\,x^2$$

Similarly, the third derivative of the function *f* can be given by any of the following:

```
diff( x^4, x, x, x);          # Short-hand format.
```

$$24\,x$$

```
diff(x^4, x$3);
```

$$24\,x$$

A few other examples of how you move from mathematical notation to *Maple* notation for derivatives is given in the following table.

Mathematical Expression	Maple *Evaluation*
$\dfrac{d(x^2 + e^{x^3})}{dx}$	`diff( x^2 + exp(x^3), x );` $2\,x + 3\,e^{x^3}x^2$
$\dfrac{d(\cos(y^2) + y^5)}{dy}$	`diff( cos(y^2) + y^5, y );` $-2\sin(y^2)\,y + 5\,y^4$

$\left.\dfrac{df}{dt}\right|_{t=1}$, where
 	
 $f(t) = t^2 + t^3 + \ln(t)$
	`f := t -> t^2 + t^3 + ln(t);`
	 `subs( t=1, diff( f(t), t ));`
	
	 6

In the last example above, we can also calculate $f'(1)$ quickly by using the **D** command. In *Maple* the command **D(f)** means f', the derivative of *f*.

```
f := t -> t^2 + t^3 + ln(t);
```

```
D(f)(t);                      # This calculates f'(t).
```

$$2t + 3t^2 + \frac{1}{t}$$

```
D(f)(1);                      # This calculates f'(1).
```

$$6$$

Differentiation Rules

Maple knows all the formal computation rules in differentiation, such as:

```
f := 'f'; g := 'g';          # Clears any previous definition of f and g.
```

```
diff(f(x)*g(x), x);          # The product rule
```

$$\left(\frac{\partial}{\partial x} f(x)\right) g(x) + f(x) \left(\frac{\partial}{\partial x} g(x)\right)$$

```
diff(f(x)/g(x), x);          # The quotient rule.
```

$$\frac{\frac{\partial}{\partial x} f(x)}{g(x)} - \frac{f(x)\left(\frac{\partial}{\partial x} g(x)\right)}{g(x)^2}$$

```
diff(f(x)^n, x);             # The power rule.
```

$$\frac{f(x)^n \, n \left(\frac{\partial}{\partial x} f(x)\right)}{f(x)}$$

How about the rule for differentiating the product of three functions?

```
diff( f(x)*g(x)*h(x), x);
```

$$\left(\frac{\partial}{\partial x} f(x)\right) g(x)\, h(x) + f(x) \left(\frac{\partial}{\partial x} g(x)\right) h(x) + f(x)\, g(x) \left(\frac{\partial}{\partial x} h(x)\right)$$

More Examples

Definition of Derivative

■ **Example.** Consider the function

```
f := x -> x^2*sin(x) + cos(x);
```

According to the definition of derivative, we have $f'(x) = \lim\limits_{h \to 0} (f(x+h) - f(x))/h$.

So when h is sufficiently small, say $h = 0.1$, we would expect $(f(x+0.1) - f(x))/0.1$ to be very close to $f'(x)$. We can see this by plotting these two functions on the same graph:

```
plot({diff(f(x),x), (f(x+0.1)-f(x))/0.1}, x=-3..3);
```

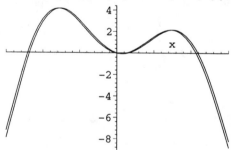

The result can be even better if we choose $h = 0.01$:

```
plot({diff(f(x),x), (f(x+0.01)-f(x))/0.01},
    x=-3..3);
```

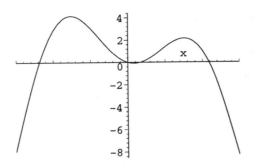

The Geometry of the Derivative

■ **Example.** Consider

```
f := x -> 3*x^2 - 6*x*cos(x);
```

Then the value of the derivative at $x = a$, $f'(a)$, gives the slope of the tangent line at that point. Using this result, the equation of the tangent line is given by:

$$y = f(a) + f'(a)(x - a).$$

We can check this by combining the graph of $y = f(x)$ with the tangent lines at $x = -2.7$, $x = -1.4$ and $x = 0.9$. (Recall that **D(f)** means f'.)

```
graph1 := plot(f(x), x=-3.5..2, thickness=3):

line1 := plot(f(-2.7)+D(f)(-2.7)*(x+2.7),
              x=-3.5..-2):
line2 := plot(f(-1.4)+D(f)(-1.4)*(x+1.4), x=-2..0):
line3 := plot(f(0.90)+D(f)(0.90)*(x-0.90), x=0..2):

with(plots):
display({graph1, line1, line2, line3});
```

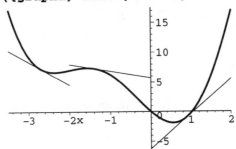

Troubleshooting Q & A

Question... *Maple* returns my limit expression unevaluated. What does that mean?

Answer... This means that *Maple* cannot determine the value of the limit. You may be able to see what the limit is by looking at the graph of the function. (Recheck the example we gave at the start of this chapter.)

Question... I keep getting zero for the derivative of a non constant function. What should I check?

Answer... You should check the following:

- Make sure that you are differentiating with respect to the correct variable. For example, you may have defined a function in terms of t but differentiated with respect to x.

- A common mistake is to type **diff(f, x);** to find the derivative, df/dx. This is wrong, because this command means the differentiation of the *variable f* and not the expression $f(x)$. You have to type **diff(f(x), x);**.

Question... I cannot assign a new function which depends on the **diff** command. What should I look for?

Answer... The usual way of defining a new function using **:=** doesn't work very well with the **diff** command. For example, if you try

```
g := x -> diff(x^2, x) ;
g(x);
```

$$2\,x$$

It looks OK, but when you ask, say, for the value of g at $x = 2$

```
g(2);
Error, (in g) wrong number (or type) of parameters in
function diff
```

This is wrong because *Maple* is trying to calculate g(2) = diff(2^2, 2) which does not make sense. The correct way is to define g using the **D(f)** command.

```
f := x -> x^2;
g := x -> D(f)(x);
```

$$g = x \rightarrow 2x$$

```
g(2);
```

$$4$$

Another method is to use the **unapply** command:

```
g := 'g';               # Clear the previous definition of g.
```

$$g := g$$

```
g := unapply( diff(x^2, x), x);
```

$$g = x \rightarrow 2x$$

```
g(2);
```

$$4$$

CHAPTER 12
Integration

Antidifferentiation

The Integrate Command

You can use the **int** command to compute $\int f(x)\,dx$. It has the form:

```
int( f(x) , x );
```

You specify a function or an expression to integrate, as well as the variable in which the integration is to take place.

For example, to compute $\int x^2\,dx$, use this syntax:

```
int(x^2, x);
```

$$\frac{1}{3}x^3$$

> **Note:** *Maple* does not put the "+ C" in the answer of integration.

Maple can integrate almost every integral that can be done using standard integration methods (e.g., substitution, integration by parts, partial fractions). Here are some typical integrations:

Integral	*In* Maple	*Comment*
$\int x\ln(x)\,dx$	`int(x*ln(x), x);` $$\frac{1}{2}x^2\ln(x)-\frac{1}{4}x^2$$	This one uses integration by parts!
$\int\dfrac{y^2}{\sqrt{1-y^2}}\,dy$	`int(y^2/sqrt(1-y^2), y);` $$-\frac{1}{2}y\sqrt{1-y^2}+\frac{1}{2}\sin^{-1}(y)$$	This is computed using a trigonometric substitution (write $y=\sin u$).
$\int\sin(\cos(y^2))\,dy$	`int(sin(cos(y^2)), y);` $$\int\sin(\cos(y^2))\,dy$$	This integrand has no closed-form antiderivative.

When *Maple* can't handle an integral, it usually returns your input unevaluated. This can mean either that it's not possible to find an antiderivative in closed form or that *Maple* hasn't yet been programmed to do the integral.

Definite Integrals

A definite integral $\int_a^b f(x)\,dx$ is computed in *Maple* with this form of the **int** command:

```
int( f(x) , x = a..b );
```

Maple will try to find an antiderivative first, then evaluate it at the endpoints and subtract (according to the Fundamental Theorem of Calculus). Here are some examples:

Integral	*In* Maple	Comment
$\int_1^2 x^2\,dx$	`int(x^2, x=1..2);` $$\frac{7}{3}$$	$\left.\dfrac{x^3}{3}\right|_1^2 = \dfrac{8}{3} - \dfrac{1}{3} = \dfrac{7}{3}$
$\int_2^\infty \dfrac{1}{5+t^2}\,dt$	`int(1/(5+t^2), t=2..infinity);` $$\frac{1}{10}\sqrt{5}\pi - \frac{1}{5}\sqrt{5}\tan^{-1}\!\left(\frac{2}{5}\sqrt{5}\right)$$	This is an improper integral that involves $+\infty$.
$\int_0^1 \sqrt{\cos(x^2)}\,dx$	`int(sqrt(cos(x^2)), x=0..1);` $$\int_0^1 \sqrt{\cos(x^2)}\,dx$$	There's no antiderivative. *Maple* can't evaluate it at the endpoints.

Numerical Integration

The evalf and int Commands

We can use the **evalf** and **int** commands together to find a numerical approximation for the integral $\int_a^b f(x)\,dx$ with the following syntax:

```
evalf( int( f(x) , x = a ..b ));
```

This combination (**evalf** and **int**) can calculate almost all definite integrals, including $\int_0^1 \sqrt{\cos(x^2)}\,dx$ for which **int** failed (as you saw earlier). Also, the computation is quick.

■ **Example.** To get approximate, numerical values for the integrals $\int_0^2 x^2\,dx$ and $\int_0^1 \sqrt{\cos(x^2)}\,dx$, use these commands:

```
evalf(int(x^2, x=0..2));
        2.666666667
```
```
evalf(int(sqrt(cos(x^2)), x=0..1));
        .9485216211
```

More Examples

Area Between Curves

■ **Example.** To approximate the area bounded by the curves $p(x) = x^5 - 20x^3$ and $q(x) = 30 - x^5$, we start by sketching the curves. First, let's see where they intersect:

```
p := x -> x^5 - 20*x^3;
q := x -> 30 - x^5;

fsolve( p(x) = q(x), x);
```
$$-3.080038835,\ -1.206322273,\ 3.231778500$$

The values -3.08, -1.206 and 3.232 give approximations for the x-coordinates of the intersections. We can see the area between these two curves in the interval $-3.1 \le x \le 3.3$:

```
plot({p(x),q(x)}, x=-3.1..3.3);
```

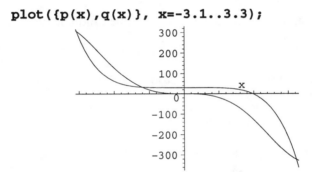

The area can be approximated with $\int_{-3.08}^{3.232} |p(x) - q(x)|\, dx$ using **evalf** and **int**:

```
evalf( int( abs( p(x)-q(x)), x=-3.08..3.232));
```
$$388.8535522 \qquad \text{\# You will get this result from R3 or R5.}$$

Comment: There is a bug in *Maple V* Release 4 that gives inaccurate answers to certain numerical integrations which involve the absolute value function. Release 4 will give you a *negative* answer in the above numerical integration! This problem has been fixed in Release 5. Nevertheless, you can get a correct answer in Release 4 by writing $\int_{-3.08}^{3.232} |p(x) - q(x)|\, dx$ as $\int_{-3.08}^{-1.206} p(x) - q(x)\, dx + \int_{-1.206}^{3.232} q(x) - p(x)\, dx$:

```
int(p(x)-q(x), x=-3.08..-1.206) +
    int( q(x)-p(x), x=-1.206..3.232);
```
$$388.8535223 \qquad \text{\# R4 now gives you a right answer.}$$

Fundamental Theorem of Calculus

■ **Example.** The Fundamental Theorem of Calculus states that if f is a continuous function, then $\dfrac{d}{dx}\int f(x)\,dx = f(x)$. So if you integrate a function f and then differentiate the result, you should expect to get f back again. Let's try it:

```
int( 1/(x^3+1), x);
```

$$\frac{1}{3}\ln(1+x) - \frac{1}{6}\ln(x^2 - x + 1) + \frac{1}{3}\sqrt{3}\tan^{-1}\left(\frac{1}{3}\sqrt{3}(2x - 1)\right)$$

diff(", x);   # or **diff(%, x);** for R5.

$$\frac{1}{3(1+x)} - \frac{1}{6}\frac{(2x-1)}{(x^2-x+1)} + \frac{2}{3+(2x-1)^2}$$

It looks like *Maple* messed up! But wait . . .

simplify(");   # or **simplify(%);** for R5.

$$\frac{1}{(x+1)(x^2-x+1)}$$

which is of course the same as $\frac{1}{(x^3+1)}$.

Plotting an Antiderivative

■ **Example.** *Maple* cannot find an explicit formula for the antiderivative

$$\int 100\ln(x)e^{(-x^2)}\ dx.$$

f := x->100*ln(x)*exp(-x^2):
int(f(x),x);

$$\int 100\ln(x)e^{(-x^2)}\ dx$$

However, you can still plot its graph with the help of the **evalf** and **int** commands. The Fundamental Theorem of Calculus says that $F(x) = \int_1^x 100\ln(t)e^{(-t^2)}\ dt$ is an

antiderivative of $100\ln(x)e^{(-x^2)}$. *Maple* can compute values of F using numerical integration (**evalf** and **int**) and you then can **plot** these values.

plot(evalf(int(f(t), t=1..x)), x=1..3);

Useful Tips

 Always use **evalf** and **int** to evaluate a definite integral, unless you need an exact answer. In many cases, **int** will neither work nor give you a useful result. Even when **int** does work, it can be slow.

Troubleshooting Q & A

Question... I got some strange answers for definite integrals that used names like "*erf*," "*fresnelS*," and "*fresnelC*." What happened?

Answer... *Maple* knows about a lot of **special functions** that appear quite often in integration problems. For example, "erf" is the *Maple* name for the error function

$$erf(x) = \frac{2}{\sqrt{\pi}} \int_0^x e^{-t^2} dt$$

This is an integral that does not have a closed-form antiderivative, but its values are well known.

You might see something like this:

```
int(exp(-t^2), t=0..1);
```
$$\frac{1}{2}\sqrt{\pi}\; erf(1)$$

```
evalf(");                # or evalf(%); for R5
```
$$.7468241330$$

Maple has reported $\int_0^1 e^{-t^2} dt = \frac{\sqrt{\pi}}{2}(\frac{2}{\sqrt{\pi}}\int_0^1 e^{-t^2} dt) = (\frac{\sqrt{\pi}}{2})erf(1) \approx 0.747.$

Question... When I tried a definite integral, I got an error message "Error, ... cannot evaluate boolean." What went wrong?

Answer... If your definite integral contains undefined variables, *Maple V* Release 3 (or earlier) will have trouble with them. For example,

```
int( sqrt(x+a), x=0..1 );     # In Release 3 or earlier.
Error, (in int/cook/IIntd1c) cannot evaluate boolean
```

Nevertheless, *Maple* knows how to find its indefinite integral:

```
int( sqrt(x+a), x);
```
$$\frac{2}{3}(x+a)^{3/2}$$

CHAPTER 13
Series and Taylor Series

Series

The sum Command

You can use the **sum** command to add up a finite number of terms of an indexed expression. To find $\sum_{n=n_0}^{n_1} expression$, you type:

> **sum(** *expression* **, n =** n_0 **..** n_1 **);**

For example,

> **sum(n^2, n=1..20);**   # Computes $\sum_{n=1}^{20} n^2$.
>
>   2870

> **sum(sin(n)/n, n=1..5);**
>
> $$\sin(1) + \frac{\sin(2)}{2} + \frac{\sin(3)}{3} + \frac{\sin(4)}{4} + \frac{\sin(5)}{5}$$

The **sum** command has moderate success even with symbolic summations. For example, $\sum_{n=0}^{k} r^n = \frac{1-r^{k+1}}{1-r}$ is a partial sum for a geometric series:

> **sum(r^n, n=0..k);**
>
> $$\frac{r^{k+1}}{r-1} - \frac{1}{r-1}$$

Numerical Summation

If you want an approximate value for a summation, use **evalf** together with **sum**. It has the syntax:

> **evalf(sum(** *expression* **, n =** n_0 **..** n_1 **));**

For example,

> **evalf(sum(1/n^2, n=1..20));**
>
>   1.596163244

> **evalf(sum(1/n^2, n=1..1000));**
>
>   1.643934568

Infinite Series

The **sum** command can also be used to find an approximate value of certain infinite series. For example, $\sum_{n=1}^{\infty} \frac{1}{n^2}$ is known to converge:

```
sum( 1/n^2, n = 1..infinity) ;
```

$$\frac{\pi^2}{6}$$

On the other hand, the harmonic series $\sum_{n=1}^{\infty} \frac{1}{n}$ diverges:

```
sum( 1/n, n = 1..infinity) ;
```

$$\infty$$

Some Famous Infinite Series

■ **Example.** Here are some well-known infinite series:

$$\sum_{n=0}^{\infty} \frac{1}{n!} = e, \qquad \sum_{n=0}^{\infty} \frac{1}{(2n+1)^2} = \frac{\pi^2}{8}, \quad \text{and} \quad \sum_{k=1}^{\infty} \frac{k}{e^{2\pi k} - 1} = \frac{1}{24} - \frac{1}{8\pi}$$

Maple recognizes the first two symbolically but can only compute the third numerically.

```
sum(1/n!, n=0..infinity);
```

$$e$$

```
sum(1/(2*n+1)^2, n=0..infinity);
```

$$\frac{\pi^2}{8}$$

```
evalf( sum( k/(exp(2*Pi*k) -1), k=1..infinity));
```

$$.001877930894$$

```
evalf(1/24 - 1/(8*Pi));
```

$$.00187793091$$

Taylor Series

Taylor Polynomials

Recall that if a function f satisfies certain reasonable conditions, then it can be approximated by a polynomial $p_n(x)$ of degree n near a point $x = a$ defined by:

$$p_n(x) = f(a) + \frac{f'(a)}{1!}(x-a) + \frac{f''(a)}{2!}(x-a)^2 + \cdots + \frac{f^{(n)}(a)}{n!}(x-a)^n$$

The polynomial $p_n(x)$ is called the Taylor polynomial of f of degree n about $x = a$.

We can use the **sum** command to write out Taylor polynomials explicitly. For example, e^{2x} has the following sixth-degree Taylor polynomial about the origin:

```
f := x -> exp(2*x);
f(0)+ subs(t=0, sum(diff(f(t), t$k)*x^k/k!, k=1..6));
```

$$1 + 2\,e^0\,x + 2\,e^0\,x^2 + \frac{4}{3}\,e^0\,x^3 + \frac{2}{3}\,e^0\,x^4 + \frac{4}{15}\,e^0\,x^5 + \frac{4}{45}\,e^0\,x^6$$

The taylor and convert Commands

Instead of using the clumsy expression shown above, you can use the **taylor** and **convert** commands as follows to produce the Taylor polynomial of degree $n-1$ about $x = a$:

> **convert(taylor(** *function,* **x = a, n), polynom);**

For example,

> **convert(taylor(cos(x), x = 0, 3), polynom);**

$$1 - \frac{x^2}{2}$$

> **convert(taylor(cos(x), x = 0, 8), polynom);**

$$1 - \frac{x^2}{2} + \frac{x^4}{24} - \frac{x^6}{720}$$

> **convert(taylor(cos(x), x = 0.5, 8), polynom);**

$$1.117295331 - .4794255386x - .4387912810(x - .5)^2$$
$$+ .07990425645(x - .5)^3 + .03656594008(x - .5)^4$$
$$- .003995212822(x - .5)^5 - .001218864669(x - .5)^6$$
$$+ .00009512411480(x - .5)^7$$

The **taylor** command by itself actually gives a Taylor polynomial together with a remainder term of degree n, $O\left((x-a)^n\right)$.

> **taylor(cos(x), x = 0, 8);**

$$1 - \frac{x^2}{2} + \frac{x^4}{24} - \frac{x^6}{720} + O\left(x^8\right)$$

Applying **convert** with the **polynom** option to this result will drop off the remainder term and give the Taylor polynomial of degree $n-1$.

> **convert(", polynom);**   # or **convert(%, polynom);** in R5.

$$1 - \frac{x^2}{2} + \frac{x^4}{24} - \frac{x^6}{720}$$

More Examples

Compare Graphically a Function with Its Taylor Polynomials

■ **Example.** Consider the function $f(x) = e^{-0.3x} + \sin(x)$:

```
f := x -> exp(-0.3*x) + sin(x);
pict1 :=  plot(f(x), x=-3..3, thickness = 3) :
```

Let us compare f graphically with some of its Taylor polynomials, say near $x = 1$. We start with the Taylor polynomial of degree 3 (but we will not show you the picture right away).

```
pict3 := plot( convert(taylor(f(x), x=1, 4), polynom),
                x=-3..3):
```

> **Note:** We have to use the number **4** in the command to give us the third degree polynomial.

`display({pict1,pict3});`   # The graph of *f* is thicker in the picture.

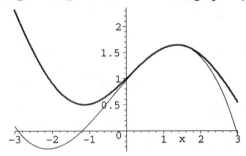

Next we increase the degree of the Taylor polynomial to 5 and then 9:

```
pict5 := plot( convert(taylor(f(x), x=1, 6),polynom),
                x=-3..3):
display({pict1,pict5});
```

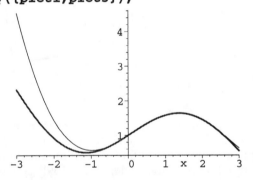

```
pict9 := plot(convert(taylor(f(x), x=1, 10),polynom),
                x=-3..3 ):
display({pict1,pict9});
```

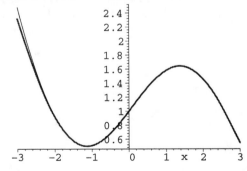

We can see that the approximation improves as we increase the degree of the Taylor polynomial.

Interval of Convergence

■ **Example.** Consider the function $g(x) = \dfrac{1}{1+x^2}$:

```
g := x -> 1/(1+x^2):
```

The theory of infinite series tells us that the Taylor series for $g(x)$ near x = 1 only converges on a certain interval centered at 1. We would like to see this result with *Maple*. We first set up graphs of the function g and its of Taylor polynomial approximations.

```
gPict :=  plot(g(x), x=-2..4, y=-2..5, thickness=3):
gPict4 := plot( convert(taylor(g(x), x=1, 4),polynom),
               x=-2..4, y=-2..5, color = blue):
gPict10:= plot( convert(taylor(g(x),x=1, 10),polynom),
               x=-2..4, y=-2..5, color = red):
gPict20:= plot( convert(taylor(g(x), x=1,20),polynom),
               x=-2..4, y=-2..5, color = green):
```

Now we use the display command to see the graphs together.

```
display({gPict, gPict4, gPict10, gPict20});
```

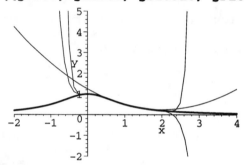

Notice that increasing the degree of the Taylor polynomial improves the approximation on a certain interval around 1. However, the polynomials turn away sharply at points outside this interval of convergence, say at $x = 2.2$. This cannot be improved no matter the degree of the Taylor polynomial.

```
gPict100:= plot(
          convert(taylor(g(x), x=1,100),polynom),
          x=-2..4, y=-2..5, color = blue):
display({gPict, gPict100});
```

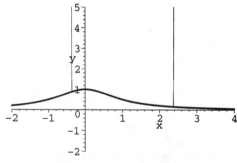

We can tell from the picture that the taylor polynomial appears to converge to g only in the interval $-0.2 \le x \le 2.2$. One can compute theoretically that the interval of convergence is $1 - \sqrt{2} \le x \le 1 + \sqrt{2}$.

Troubleshooting Q & A

Question...   I got some strange answers when summing a series that used names like "Psi" or "Ψ." What happened?

Answer...   These are **special functions** that appear quite often in summation of a series. Their values are well known. For example, you might see something like this:

```
sum( 1/n^2, n = 1..1000) ;
```

$$-\Psi(1,\ 1001) + \frac{1}{6}\pi^2$$

```
evalf(");        # or evalf(%); in R5.
```
              1.643934568

Question...   When I calculated an infinite series using **evalf** and **sum**, *Maple* returned the answer "FAIL(...)." Does this mean that the series diverges?

Answer...   No. This only means that *Maple* failed to give you an answer. For example, $\sum_{n=1}^{\infty} \frac{\cos(n)}{n^2}$ converges while $\sum_{n=1}^{\infty} \cos(n)$ diverges, but both will get a similar response from *Maple* V Release 3 or earlier, as shown below:

```
evalf(sum( cos(n)/n^2, n=1..infinity));
FAIL(.3174253143)
```

```
evalf(sum( cos(n), n=1..infinity));
FAIL(-1.330601313)
```


Question...   I found a Taylor series using the **taylor** command, but I had trouble plotting it. What should I check for?

Answer...   The **taylor** command gives both a Taylor polynomial and a formal error term $O\left((x-a)^n\right)$. The **plot** command does not work because of this term. You have to use **convert** to drop this term before you can **plot** it. For example:

```
plot( taylor( sin(x), x=0, 5), x=-3..3);
Plotting error, empty plot
```

Instead, you should use

```
plot( convert(taylor( sin(x), x=0, 5), polynom),
          x=-3..3);
```


Question...   When I tried plotting a function and a Taylor polynomial on the same graph, the function disappeared. What happened?

```
plot({sin(x),
        convert(taylor(sin(x),x=0, 10), polynom)},
        x= -10..10);
```

Answer... First check the plots separately. If they both work, look at the scales on the y-axis. If the scales are very different, one of the graphs may appear as a straight line over the *x*-axis when you put the 2 curves together. Look at the scale on the *y*-axis. The large *y* scale will make most interesting features of your original function disappear. Specify a range for the *y*-axis

```
plot({sin(x),
      convert(taylor(sin(x),x=0, 10), polynom)},
      x= -10..10, y= -2..2);
```

CHAPTER 14
Solving Differential Equations

Symbolic Solutions of Equations

The dsolve Command

Using the **dsolve** command, you can symbolically solve ordinary differential equations that involve $y(x)$ and x. This command is used in the form:

dsolve(*the differential equation* **, y(x));**

> **Note:** The differential equation is entered using the equal sign = and the y must appear as **y(x)**.

Here are some simple examples.

Equation	To Solve It in Maple	Comment
$y' = 3x^2 y$	`dsolve( diff(y(x), x ) = 3*x^2*y(x), y(x));` $$y(x) = \_C1\ e^{(x^3)}$$	**diff(y(x), x)** is used to denote y' in *Maple*.
$y\,y' = -x$	`dsolve(y(x)*diff(y(x),x) = -x, y(x));` $$y(x) = \sqrt{-x^2 + \_C1}\,, \quad y(x) = -\sqrt{-x^2 + \_C1}$$	There are two solutions for this equation.
$y'' + y' - y = 0$	`dsolve( diff(y(x), x$2) + diff(y(x),x) - y(x) = 0, y(x));` $$y(x) = \_C1\ e^{\left(-\frac{1}{2}(\sqrt{5}+1)x\right)} + \_C2\ e^{\left(\frac{1}{2}(\sqrt{5}-1)x\right)}$$	This is a second-order equation. We use **diff(y(x), x$2)** to denote y''.
$y' = -\cos(xy)$	`dsolve( diff(y(x),x) = -cos(x*y(x)), y(x));` (no output from *Maple*)	*Maple* will not return an output if it cannot solve the differential equation.

> **Note:** $\_C1$ and $\_C2$ denote arbitrary constants in the solutions above.

Equations with Initial or Boundary Conditions

Sometimes you need to solve a differential equation subject to initial or boundary conditions. In such a case, the format of the **dsolve** command is:

dsolve({ *a differential equation* **,** *initial or boundary condition(s)* **},** **y(x));**

Here are some examples:

Equation & Condition(s)	To Solve It in Maple
$y' = -36x$, with $y(0) = 2$	`dsolve( {diff(y(x),x) = -36*x, y(0)=2}, y(x));` $y(x) = -18\,x^2 + 2$
$y' = 3x^2 y$ with $y(1) = -1$	`dsolve({diff(y(x),x) = 3*x^2*y(x),y(1)=-1},y(x));` $y(x) = -\dfrac{e^{(x^3)}}{e}$

$y'' + y = 20\cos(x)$ with $y(0) = 0$, $y(\pi/2) = 1$	`dsolve( {diff(y(x),x$2) +y(x) = 20*cos(x),`
	 `        y(0)=0, y(Pi/2) =1}, y(x)):`
	 `simplify(");   # or simplify(%); in R5.`
	
	 $y(x) = 10\,\sin(x)\,x - 5\,\sin(x)\,\pi + \sin(x)$

$y'' - 2y' + y = x$ with $y(0) = 0$, $y'(0) = 2$	`dsolve( {diff(y(x), x$2)-2*diff(y(x),x) + y(x)`
	 `        = x, y(0)=0, D(y)(0)=2}, y(x));`
	
	 $y(x) = 2 + x - 2\,e^x + 3\,e^x\,x$

> **Note:** The initial condition $y'(0) = 2$ in the last example above is entered as **D(y)(0)=2**.

Notice that in the last two examples above, two initial or boundary conditions are needed to guarantee a unique solution for a second-order differential equation.

Numerical Solutions of Equations

The numeric Option for dsolve

By adding the **numeric** option in the **dsolve** command, you can find a numerical approximation to a solution of a differential equation, subject to given initial conditions. You use it in the following format:

> **dsolve(**{ *differential equation, initial condition(s)*}**, y(x), numeric);**

For example, you can find a numerical approximation for the solution to the equation $y' = -xy$, subject to the initial condition $y(0) = 1$ with:

> **soln := dsolve({diff(y(x),x) = -x*y(x), y(0)=1},**
 **   y(x), numeric);**
>
> *soln* := **proc**(*rkf45_x*) ... **end**

This output looks strange. Don't worry! *Maple* reports the answer as an algorithm to approximate the solution numerically.

You can now compute, for example, the values $y(0)$, $y(0.25)$, $y(0.5)$ and $y(1)$ with:

> **[soln(0), soln(0.25), soln(0.5), soln(1)];**
>
> $[[x = 0, y(x) = 1.]$, $[x = .25, y(x) = .9692331735804627]$,
 $[x = .5, y(x) = .8824968468192492]$, $[x = 1, y(x) = .6065306263304814]]$

Or you can see a graph of the solution with the **odeplot** command that's defined in the **plots** library:

```
with(plots):
odeplot(soln, [x, y(x)], 0..2);
```

In general, the **odeplot** command can show the graph of the solution $y(x)$ for $a \le x \le b$ with the syntax:

```
with(plots):
odeplot( name of the result from dsolve, [x, y(x)],
              a..b );
```

Systems of Differential Equations

Solving a System Symbolically

If you have a system of differential equations involving functions $x(t)$ and $y(t)$, *Maple* can solve the system symbolically using the format:

```
dsolve[ { system of differential equations }, {x(t), y(t)} );
```

For example, to solve for functions $x(t)$ and $y(t)$ in the system $x'(t) = x(t) - y(t)$ and $y'(t) = y(t)$, use:

```
dsolve({diff(x(t),t) = x(t) - y(t),
              diff(y(t),t) = y(t)}, {x(t), y(t)});
```

$$\{y(t) = e^t \_C2, \quad x(t) = -e^t(-\_C1 + t\_C2)\}$$

Solving a System Numerically

The **dsolve** command can solve systems of differential equations numerically using the **numeric** option, similar to what we did earlier. For example, to solve the system $x'(t) = -2x(t)^2 - y(t)$ and $y'(t) = x(t) - y(t)$ numerically with initial conditions $x(0) = 0.2$ and $y(0) = 0.1$, use:

```
soln := dsolve({diff(x(t),t) = -2*x(t)^2 - y(t),
     diff(y(t),t) = x(t)-y(t), x(0) = 0.2, y(0) = 0.1},
     {x(t), y(t)}, numeric);
```

$$soln := \mathbf{proc}(rkf45\_x) \dots \mathbf{end}$$

Again, the output is given by an algorithm. Now, for example, you can find the values of $(x(0), y(0))$ and $(x(5), y(5))$ with:

```
[soln(0), soln(5)];
```

$$[[t = 0, x(t) = .200000000000, y(t) = .10000000000],$$
$$[t = 5, x(t) = -.003731000641, y(t) = -.0154401944]]$$

You can also see the solution curve $(x(t), y(t))$ (called a phase diagram):

```
odeplot(soln, [x(t),y(t)], 0..20, numpoints = 200);
```

(We draw the curve $(x(t), y(t))$ for $0 \le t \le 20$. Also we increase the number of points to 200 to get a smoother picture.)

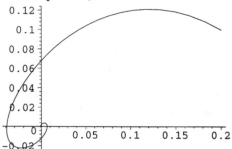

More Examples

Geometry of a First-Order Differential Equation

■ **Example.** Consider the differential equation $y' = 3x^2y$. Let's find the general solution:

```
dsolve( diff(y(x), x ) = 3*x^2*y(x), y(x));
```

$$y(x) = \_C1\, e^{(x^3)}$$

You can sketch a few particular solutions of this equation, for example, with $C1$ having values 0.5, –0.71, and 0.92. (We won't show the output here – you will see the picture later anyway):

```
curves := plot({0.5*exp(x^3), -0.71*exp(x^3),
          0.92*exp(x^3)}, x=-2..1, thickness = 3):
```

The slope field of this differential equation can be seen with the following command. We will explain this command in detail in Chapter 20 when we discuss vector fields.

```
with(plots):
pict1 := fieldplot([1, 3*x^2*y], x=-2..2, y=-2..2):
```

Now you see the solutions together with the slope field:

```
display( {pict1, curves} );
```

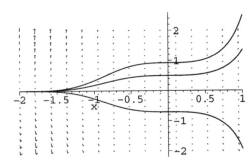

As you see from the picture, the solution curves are tangent to the vectors that make up the slope field. You may wonder why.

Recall that y' gives the slope of the graph of y at any point. Since each of the solution curves satisfies $y' = 3x^2y$, the tangent slope at each point will be $3x^2y$. At the same time, each of the vectors $(1, 3x^2y)$ in the slope field has slope $3x^2y$. This demonstrates that the solution curves are tangent to the slope field, as seen in the picture.

Useful Tips

There is a library package called **DEtools** which includes many useful commands in solving ordinary or partial differential equations. You can load the package using **with(DEtools);** and experiment with the commands.

Troubleshooting Q & A

Question... I got an error message from **dsolve**. What should I look for?

Answer... This most likely means that you did not enter the differential equation(s) correctly. The three most common mistakes made are these:

- You did not match (*parentheses*) correctly.

- You did not follow the rules: y has to be typed as **y(x)**, y' is typed as **diff(y(x), x)**, y'' is typed as **diff(y(x), x$2)**, $y'(a) = b$ is typed as **D(y)(a) = b**, and so on.

- You assigned values earlier to either the independent variable (**x**) or the function name (**y**). Try **unassign('x', 'y');** and reexecute.

Question... When I used the **numeric** option for **dsolve,** I got an error message about ". . . initial conditions." What does this mean, and what should I do?

Answer... Make sure you gave the right number of initial conditions in your input. A first-order differential equation needs one initial condition, a second-order differential equation needs two initial conditions, and so on.

CHAPTER 15

Making Graphs in Space

Graphing Functions of Two Variables

The plot3d Command

The easiest way to sketch a surface in three dimensions (in 3-D) is to use the **plot3d** command.

You input an expression that gives the height of a surface above the xy-plane, in terms of the independent variables x and y. You must also specify bounds for the x- and y-variables as $x_0 \le x \le x_1$ and $y_0 \le y \le y_1$. The **plot3d** command then has the form:

> **plot3d(** *an expression of x and y,* **x** = x_0 **..** x_1, **y** = y_0 **..** y_1**)** ;

For example, the surface whose height is $z = 4 - x^2 - y^2$ above the xy-plane, over the rectangle $-2 \le x \le 2$ and $-2 \le y \le 2$, is seen with:

```
plot3d( 4-x^2-y^2, x = -2..2, y = -2..2);
```

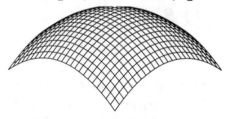

Mathematically speaking, this generates the surface $z = 4 - x^2 - y^2$ for $-2 \le x \le 2$ and $-2 \le y \le 2$. In the language of multivariable calculus, this means that **plot3d** will show you the graph of a two-variable function.

■ **Example.** The graph of the function $f(x, y) = x^2 - y^2$ looks like a saddle.

```
f := (x,y) -> x^2 - y^2 ;
plot3d( f(x,y), x = -3..3 , y = -3..3);
```

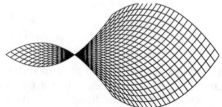

Changing the 3-D View for a Graphically Based System

In Release 5, click and hold the (left) mouse button with the cursor/arrow at any point inside the 3-D picture. As you drag the mouse with the button held down, the picture rotates. This allows you to see the picture from any viewpoint.

In Releases 3 and 4, when you drag the mouse with the button held down, a frame box rotates instead. When you release the (left) button, you can ask *Maple* to redraw the picture from this new viewpoint simply by clicking the **plot** or **R** button. (The **plot** or **R** button is available from the plot3d options tool bar or on top of the picture, depending on the type of machine and the version of *Maple* you are using.) This is what the plot3d option bar looks like for Release 4 on a Mac:

The grid Option

You can use the **grid** option to generate a more detailed picture. Consider:

```
plot3d( sin(3*y)+cos(5*x), x = -Pi ..Pi, y=-Pi..Pi);
```

The picture does not look very good. But if we add the option **grid = [50, 50]**, we'll see the graph shown with values sampled from a 50 × 50 grid, instead of the default 25 × 25 grid. This gives a smoother picture but takes longer to compute and draw.

```
plot3d( sin(3*y)+cos(5*x),
            x = -Pi ..Pi, y=-Pi..Pi, grid=[50,50]);
```

Other Useful Options

There are a number of options that you can use to add more "character" to a picture with **plot3d**. For example,

Option	What It Does
`axes = normal`	Shows the three axes in a normal way.
`axes = boxed`	Surrounds the picture with a box.
`labels = [`x`,`y`,`z`]`	Provides names to label each of the three axes.
`scaling = constrained`	Scales the graphic so that units in each direction have the same length.
`color = the color you want`	Draw the graphic with the specified color.

Also, you can redraw the picture by choosing various options under the **Style** menu, **Axes** menu **Color** menu and **Projection** menu on the menu bar.

■ **Example.** The sombrero has equation $f(x,y) = \sin(\sqrt{x^2 + y^2})/\sqrt{x^2 + y^2}$:

```
plot3d( sin(sqrt(x^2+y^2))/sqrt(x^2+y^2), x=-7..7,
    y=-7..7,grid =[30,30], axes = boxed, color = red,
    labels=[`x axis`,`y axis `,`z axis`]);
```

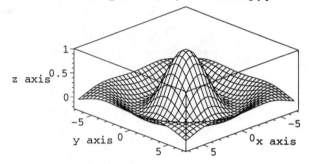

Surfaces in Cylindrical and Spherical Coordinates

Sometimes surfaces are described in terms of the cylindrical or spherical coordinate systems. *Maple* can draw such surfaces easily with the **cylinderplot** and **sphereplot** commands, respectively. These are defined in the **plots** library.

The cylinderplot Command

Points in the cylindrical coordinate system are described by quantities r, θ, and z, where

- r is the horizontal radial distance of the point from the z-axis;

- θ is the horizontal angle measured from the x-axis; and

- z is the z-coordinate in standard rectangular coordinates.

To draw the surface $r = f(\theta, z)$ for $\theta_0 \leq \theta \leq \theta_1$ and $z_0 \leq z \leq z_1$, you enter:

```
with(plots):
cylinderplot[ f(θ, z), theta = θ0..θ1, z = z0..z1);
```

> **Note:** In the **cylinderplot** command, you must enter the interval for **theta** first, then the interval for **z**. If you do not follow this order, the picture will be incorrect.

For example, to see the surface $r = z^2(\cos(3\theta))^2$, for $0 \leq \theta \leq 2\pi$ and $-2 \leq z \leq 2$:

```
with(plots):
cylinderplot( z^2*(cos(3*theta))^2, theta=0..2*Pi,
    z=-2..2);
```

Notice that this surface is very "choppy" as the horizontal angle θ varies, yet the surface is quite smooth along the vertical z-direction. This suggests that we should increase the resolution of the graphic in the θ variable (from 25 to 80) but decrease the resolution in z (from 25 to 15). We can do this with:

```
cylinderplot( z^2*(cos(3*theta))^2, theta=0..2*Pi,
        z=-2..2, grid=[80,15]);
```

The sphereplot Command

Points in the spherical coordinate system are described by quantities ρ, θ, and ϕ, where

- ρ is the radial distance in space of the point from the origin;

- θ is the horizontal angle measured from the x-axis; and

- ϕ is the vertical angle measured from the z-axis.

To draw the surface $\rho = f(\theta, \phi)$, $\theta_0 \le \theta \le \theta_1$, and $\phi_0 \le \phi \le \phi_1$, you will enter:

```
with(plots);
sphereplot(  f(θ, φ), theta = θ₀..θ₁, phi = φ₀..φ₁);
```

> **Note:** In the **sphereplot** command, you have to enter the interval for **theta** first, then the interval for **phi**. If you do not follow this order, the picture will be incorrect.

For example, to see the surface $\rho = \sqrt{\theta}\,(3 + \cos\phi)$, $0 \le \theta \le 3\pi/2$, and $0 \le \phi \le \pi$:

```
with(plots):
sphereplot( sqrt(theta)*(3 + cos(phi)),
   theta =0..3*Pi/2, phi=0..Pi);
```

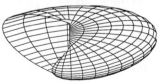

More Examples

Choosing the Right Coordinate Systems

In some cases, a surface given in rectangular coordinates will look better if you draw it using cylindrical or spherical coordinates.

■ **Example.** For example, the surface $z = \frac{x^2 - y^2}{(x^2 + y^2)^2}$ can be plotted with:

```
plot3d( (x^2-y^2)/(x^2+y^2)^2, x=-3..3, y=-3..3 );
```

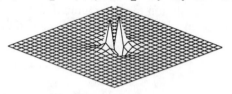

The picture is choppy especially near the origin. However, if we use cylindrical coordinates, the equation of the surface becomes

$$z = \frac{x^2 - y^2}{(x^2 + y^2)^2} = \frac{(r\cos\theta)^2 - (r\sin\theta)^2}{((r\cos\theta)^2 + (r\sin\theta)^2)^2} = \frac{r^2(\cos^2\theta - \sin^2\theta)}{r^4} = \frac{\cos 2\theta}{r^2},$$

which means $r^2 = \cos(2\theta)/z$, or equivalently $r = \sqrt{\cos(2\theta)/z}$. (We need to express the surface in the form $r = f(\theta, z)$ in order to use **cylinderplot**.)

```
top := cylinderplot( sqrt(cos(2*theta)/z),
        theta=0..2*Pi, z=0.1..2, grid =[50,15]):
bottom := cylinderplot( sqrt(cos(2*theta)/z),
        theta=0..2*Pi, z=-2..-0.1, grid =[50,15]):
display({top,bottom}, orientation=[36,77]);
```

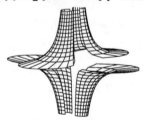

This picture is nicer!

Troubleshooting Q & A

Question... *Maple* either quit running or "crashed" when I was drawing some 3-D graphics. I also saw some messages that *Maple* "needs more memory." What should I do?

Answer... *Maple* uses a lot of memory when it creates 3-D pictures. You need a very capable system with lots of memory to run *Maple*. Most "crashes" occur because there's not enough memory on your system.

When you installed *Maple* "out of the box," your system memory parameters may not have been set high enough to sustain intensive 3-D imaging. If you can increase *Maple*'s memory partition on your system, please do so.

Question... I tried to draw a 3-D picture but got an error message telling me about a "`Plotting error.`" What happened?

Answer... This means that *Maple* has trouble using your function you specified to generate points for the picture. Check these points:

- Did you mistype the input?
- Did you use the same variables in the function as you used in specifying the intervals?
- Is the function well defined in the given intervals?

Another common mistake is to use the **plot** command to draw a 3-D picture; make sure you use the **plot3d** command.

Question... I drew the graph of a function using **plot3d**, but the picture looks different from the one shown in my calculus book. Why is that?

Answer... Two graphics may sometimes look different because they are drawn with different scales on the three axes. Adjust them with the **scaling** option. Also try different viewpoints.

Question... When I used **cylinderplot** or **sphereplot**, nothing happened. What should I do?

Answer... Did you remember to load the package **with(plots)**? Now, try the command again.

Question... The picture I got from **cylinderplot** or **sphereplot** was completely wrong. What should I check?

Answer... Check these three areas:

- Make sure you typed the input function and the intervals of the two parameters correctly.
- In **cylinderplot** you have to enter the θ-interval **theta** $= \theta_0 .. \theta_1$ first, followed by the z-interval **z** $= z_0 .. z_1$. If you enter these in the wrong order, *Maple* will reverse the sense of the variables.
- In **sphereplot** you have to enter the θ-interval **theta** $= \theta_0 .. \theta_1$ first, then the ϕ-interval **phi** $= \phi_0 .. \phi_1$. If you enter these in the wrong order, *Maple* will draw an incorrect picture.

CHAPTER 16
Level Curves and
Level Surfaces

Level Curves in the Plane

The contourplot Command

In *Maple*, the level curves (contours) of a function $f(x, y)$ are plotted with the **contourplot** command that is defined in the **plots** library. To see the level curves inside the rectangle $x_0 \le x \le x_1$, $y_0 \le y \le y_1$, you enter:

```
with(plots);
contourplot( function , x=x_0..x_1, y=y_0..y_1);
```

(This syntax looks exactly like the **plot3d** command syntax that we discussed in the previous chapter.) For example, here are some level curves of $f(x, y) = x y e^{-x^2 - y^2}$:

```
contourplot(x*y*exp(-x^2-y^2), x=-2..2, y=-2..2);
```

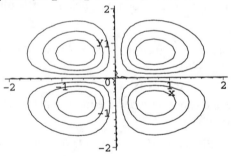

Options for ContourPlot

Some of the options we like to use with **contourplot** are the following:

Option	What It Does
contours = n	Draws n level curves
grid = [n, m]	Changes resolution of the picture
filled = true, coloring = [white, blue]	Shades the areas between level curves. Lighter blues represent lower levels, while darker blues represent higher levels.
scaling = constrained	Both the x- and y-axes are drawn with the same scale.

Here's a nicer picture than the one before:

```
contourplot(x*y*exp(-x^2-y^2), x=-2..2, y=-2..2,
    contours = 20, grid =[ 20, 20], filled=true,
    coloring=[white,blue], scaling=constrained);
```

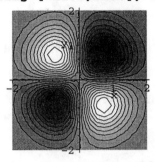

With our choice of **coloring**, lighter shades represent lower levels, while darker shades represent higher levels. We can tell from this picture that the function has its largest values near (0.8, 0.8) and (−0.8, − 0.8).

Plotting Specific Levels

You can plot specific level curves using the **contours** option in the form **contours =** [*the levels*]. The levels must be separated by commas. For example, to see the contours at levels 0, 0.1, and 0.15, without the shading:

```
contourplot(x*y*exp(-x^2-y^2), x=-2..2, y=-2..2,
    grid = [ 30,30 ], contours = [0,0.1,0.15]);
```

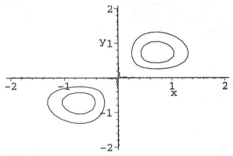

Notice that the level curve of $f(x, y) = x y e^{-x^2-y^2}$ at level 0 consists of the x- and y-axes. This is why the axes are shown in the picture.

Level Surfaces in Space

The implicitplot3d Command

If $f(x,y,z)$ is a function of three variables defined over a rectangular region $x_0 \le x \le x_1$, $y_0 \le y \le y_1$ and $z_0 \le z \le z_1$, then the level surface of f at level c can be seen with the **implicitplot3d** command. It's defined in the **plots** library.

```
with(plots);
implicitplot3d( f(x,y,z) = c, x = x_0..x_1, y = y_0..y_1,
    z = z_0..z_1);
```

You can draw more than one level surface in the same graphic, say with levels c_1

and c_2, by entering:

```
implicitplot3d({ f(x,y,z) = c₁, f(x,y,z) = c₂ },
        x = x₀..x₁, y = y₀..y₁, z = z₀..z₁);
```

Here are the level surfaces of $f(x, y, z) = x^3 - y^2 + z^2$ at the levels 1 and 10:

```
f := (x,y,z) -> x^3-y^2+z^2 ;
with(plots):
implicitplot3d({f(x,y,z)=1, f(x,y,z)=10},
        x=-2..5, y=-2..2, z=-2..3);
```

In other words, the surfaces shown above have equations $x^3 - y^2 + z^2 = 1$ and $x^3 - y^2 + z^2 = 10$.

> **Note: implicitplot3d** produces very nice graphics but unfortunately requires significant computation time.

More Examples

Comparing plot3d with contourplot

■ **Example.** Consider $f(x, y) = x^2 - y^2$. The following command will show the contours at level 0, 1, and –1.

```
contourplot(x^2- y^2, x=-2..2, y=-2..2,
    contours = [0, 1, -1], scaling=constrained) ;
```

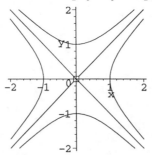

What are these curves? Recall that a contour at level c is given by the equation $f(x, y) = c$. We thus have:

- The contour at level 0 has equation $x^2 - y^2 = 0$, which consists of the two lines

$y = x$ and $y = -x$.

- The contour at level 1 has equation $x^2 - y^2 = 1$ and is a hyperbola that opens left/right.

- The contour at level -1 has equation $x^2 - y^2 = -1$ and is also a hyperbola that opens up/down.

We can see these three contours by intersecting the graph of $f(x, y) = x^2 - y^2$ by the planes $z = 1$, $z = 0$ and $z = -1$.

```
plot3d({x^2-y^2,1,0,-1},x=-2..2,y=-2..2);
```

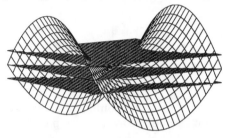

Also, you can see the relationship between level curves and the graph through interacting with *Maple*'s **contourplot3d** command (available for Release 4 or higher). Type:

```
with(plots);
contourplot3d( x^2-y^2,x=-2..2, y=-2..2,
   orientation = [0, 0], filled = true);
```

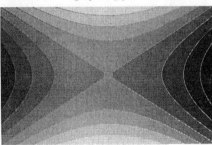

You see the level curves here. Now, click on the picture, rotate the frame box to a 3-D position, and then redraw:

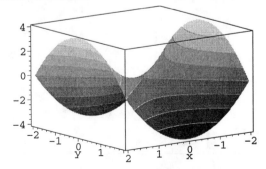

You can now see how the level curves are formed from the graph.

Curves in the Plane Defined by an Equation

■ **Example**. The equation $2x^2 - 3xy + 5y^2 - 6x + 7y = 8$ defines a rotated ellipse in the plane. We could use **implicitplot** to draw it. But it's also just the level curve of the function $f(x, y) = 2x^2 - 3xy + 5y^2 - 6x + 7y$ at level 8. We can see it with **contourplot**:

```
contourplot(2*x^2 - 3*x*y + 5*y^2 - 6*x + 7*y,
    x=-2..5, y=-3..2, contours =[8],grid = [30,30]);
```

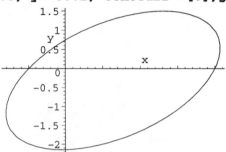

(The intervals used in the command above were arrived at after some experimentation, so that we could give you a nice picture.)

The Quadric Surfaces in Space

The Quadric Surfaces are those surfaces in space which can be given by an equation of the form:

$$Ax^2 + By^2 + Cz^2 + Dxy + Eyz + Fxz + Gx + Hy + Iz + J = 0,$$

where $A, B, C, \ldots, J$ are constants. These surfaces are discussed in detail in every multivariable calculus book. With the help of the **implicitplot3d** command, we can easily see pictures of various quadric surfaces.

■ **Example**. The equation $\dfrac{x^2}{2^2} + \dfrac{y^2}{3^2} - \dfrac{z^2}{4^2} = 1$ defines a hyperboloid of one sheet:

```
implicitplot3d(x^2/2^2 + y^2/3^2 - z^2/4^2 = 1,
    x=-10..10, y=-10..10,z=-10..10 );
```

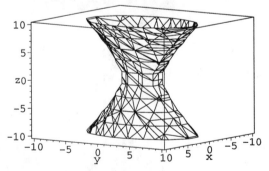

Troubleshooting Q & A

Question...   When I tried **contourplot** or **implicitplot3d**, *Maple* returned the command to me unevaluated. What should I check?

Answer... Most likely you did not load the **plots** library before using the command. Type:

```
with(plots);
```

Now, try the command again.

Question... I got the error message "Plotting error, empty plot" from **contourplot**, when I specified a list of levels in the **contours** option. What happened?

Answer... There are two likely possibilities:

- The level curves you specified do not exist (e.g., the contour with level 0 is empty for the function $f(x, y) = x^2 + y^2 + 1$).

- The level curves you wanted to see do not lie inside the rectangle you gave (e.g., the contour $x^2 + y^2 = 1$ does not lie inside the rectangle $2 \le x \le 3$, $2 \le y \le 3$).

Question... I got an empty picture from **contourplot** when I specified a list of levels in the **contours** option. What happened?

Answer... Most likely the intervals you specified lie completely inside the level curve. For example, the contour $x^2 + y^2 = 9$ contains the rectangle $-2 \le x \le 2$, $-2 \le y \le 2$, so you get an empty picture when you enter:

```
contourplot(x^2+y^2, x=-2..2, y=-2..2,
    contours = [9]);
```

Question... I got the error message "Plotting error, empty plot" from **implicitplot3d**. What happened?

Answer... There are three likely possibilities:

- The level surface does not exist, so nothing can be shown in the output (e.g., the level surface $x^2 + y^2 + z^2 = -1$).

- The level surface you're trying to see doesn't lie inside the region $x_0 \le x \le x_1$, $y_0 \le y \le y_1$, and $z_0 \le z \le z_1$ you gave. (This is similar to the problem addressed in the second question above.) Recheck both the level and the region you specified.

- You did not define a three-variable function $f(x, y, z)$ correctly.

CHAPTER 17
Partial Differentiation and Multiple Integration

Partial Derivatives

The diff Command

The **diff** command we used in Chapter 11 is actually a partial differentiation operator. It differentiates an expression with respect to a specified variable, treating all other symbols as constants.

> **diff(** *the function or expression* **,** *variable* **);**

For example, if $f(x, y) = 3xy^2 - 5y \sin x$, its partial derivatives $f_x = \dfrac{\partial f}{\partial x}$ and $f_y = \dfrac{\partial f}{\partial y}$ are computed with:

```
f := (x,y) -> 3*x*y^2 - 5*y*sin(x):
diff( f(x,y), x );
```
$$3y^2 - 5y\cos(x)$$

```
diff( f(x,y), y );
```
$$6xy - 5\sin(x)$$

You find higher-order derivatives by listing the variables in the order of differentiation. For example, $f_{xxy} = \dfrac{\partial^3 f}{\partial y \partial x^2}$ represents taking the partial derivative "first by x, then by x, then by y." You compute this with:

```
diff( f(x,y), x, x, y );
```
$$5\sin(x)$$

Double and Triple Integrals

Iterated Double Integrals

The **int** command is well suited to computing double integrals because it treats any variable that is not named in the command as a constant. The *inner* integral in $\int_a^b \int_{g_1(x)}^{g_2(x)} f(x, y)\, dy\, dx$ is evaluated as:

```
int( f(x, y) , y = g_1(x) .. g_2(x));
```

92

Then the outer integral is computed with

```
int( " , x = a..b);
```

You can do them both at once with:

```
int(int( f(x, y) , y = g₁(x)..g₂(x)), x = a..b);
```

Similarly, the iterated double integral $\int_c^d \int_{h_1(y)}^{h_2(y)} f(x, y)\, dx\, dy$ is computed with:

```
int(int( f(x, y) , x = h₁(y)..h₂(y)), y = c..d);
```

■ **Example.** The double integral $\int_0^2 \int_0^{2x} (3x^2 + (y-2)^2)\, dy\, dx$ is computed with:

```
int(int(3*x^2 + (y-2)^2, y=0..2*x), x=0..2);
```

$$\frac{88}{3}$$

The integration above takes place over the region D defined by the inequalities:

$$0 \le x \le 2 \quad \text{and} \quad 0 \le y \le 2x$$

(See the picture to the right.) D can also be described by the inequalities:

$$0 \le y \le 4 \quad \text{and} \quad y/2 \le x \le 2$$

It follows that the double integral $\int_0^2 \int_0^{2x} (3x^2 + (y-2)^2)\, dy\, dx$

has the same value as $\int_0^4 \int_{y/2}^2 (3x^2 + (y-2)^2)\, dx\, dy$:

```
int(int(3*x^2 + (y-2)^2, x = y/2..2), y = 0..4);
```

$$\frac{88}{3}$$

Iterated Triple Integrals

An iterated triple integral $\int_a^b \int_{g_1(x)}^{g_2(x)} \int_{h_1(x, y)}^{h_2(x, y)} f(x, y, z)\, dz\, dy\, dx$ is evaluated using the **int** command three times:

```
int(int(int( f(x,y,z) , z = h₁(x, y)..h₂(x, y)),
                y = g₁(x)..g₂(x)), x = a..b);
```

Other variations in the order of integration can be evaluated similarly.

For example, to evaluate $\int_{-3}^3 \int_{-\sqrt{9-x^2}}^{\sqrt{9-x^2}} \int_{x+y}^{3+y} z^2\, dz\, dy\, dx$, write:

```
int(int(int(z^2, z=x+y..3+y),
      y=-sqrt(9-x^2)..sqrt(9-x^2)), x=-3..3);
```

$$\frac{567}{4}\pi$$

Numerical Integration

If you want to find a numeric approximation for a double or triple integral, you should use **evalf** together with the **int** commands.

For example, to find $\int_{-3}^{3}\int_{-\sqrt{9-x^2}}^{\sqrt{9-x^2}}\int_{x+y}^{3+y} z^2 \, dz \, dy \, dx$ numerically, write:

```
evalf(int(int(int( z^2, z = x+y..3+y),
          y=-sqrt(9-x^2)..sqrt(9-x^2)), x=-3..3));
            445.3207587
```

As you may expect, numerical integration will give you an answer quickly in most cases, and it can even be used when **int** fails.

More Examples

Critical Points and the Hessian Test

■ **Example.** Suppose that $f(x,y) = x^4 - 3x^2 - 2y^3 + 3y + 0.5xy$. We can find its critical points by solving the equations $f_x = 0$ and $f_y = 0$ simultaneously:

```
f := (x,y) -> x^4 - 3*x^2 - 2*y^3 + 3*y + 0.5*x*y;
criticalpt := [fsolve( {diff(f(x,y),x) = 0,
                   diff(f(x,y),y)=0}, {x,y},maxsols=7)];
```

$$
\begin{aligned}
criticalpt := [&\{x = -1.250162405, y = .6291421139\}, \\
&\{x = 1.255942703, \ y = -.7776000848\}, \\
&\{x = 1.191117672, \ y = .7741187287\}, \\
&\{x = .5935572277e\text{-}1, \ y = .7105957432\}, \\
&\{x = -.5877159444e\text{-}1, \ y = -.7036351094\}, \\
&\{x = -1.197482099, \ y = -.6326213916\}]
\end{aligned}
$$

There are six critical points. We will define the discriminant

$$D = (f_{xx})(f_{yy}) - (f_{xy})^2$$

and evaluate it at each critical point. The Hessian Test you learned in multivariable calculus says:

- If $D < 0$, the critical point is a saddle point.

- If $D > 0$, the critical point is a local maximum when f_{xx} is negative.

- If $D > 0$, the critical point is a local minimum when f_{xx} is positive.

We compute the discriminant D and f_{xx} at each of the critical points:

```
dis := diff(f(x,y),x,x)*diff(f(x,y),y,y)
                   - diff(f(x,y),x,y)^2 ;
```

$$dis := -12 (12 \, x^2 - 6) \, y - .25$$

```
subs( criticalpt[1],[dis, diff(f(x,y),x,x)]);
            [-96.54552914, 12.75487247]
```

This means that the first critical point is a saddle point. We can check all the critical points at once with:

```
seq( subs(criticalpt[n], [dis, diff(f(x,y),x,x)]),
    n=1..6);
```

[–96.54552914,   12.75487247],
[120.3903441,   12.92870488],
[–102.6671685,   11.02513571],
[50.55238934,   –5.957722778],
[–50.56174649, -5.958550796],
[84.83171040,   11.20756052]

This result shows that the discriminant is positive at the second, fourth, and sixth of the critical points. Since f_{xx} is positive at the second and sixth, those critical points [(1.256, –0.778) and (–1.197, –0.633)] will be local minima for f. Also, the point (0.059, 0.711) will be a local maximum.

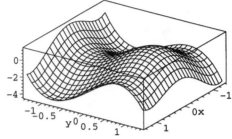

The three remaining critical points have negative discriminants, so each is a saddle point for f. You can check these points from the graph of f shown above.

Method of Lagrange Multipliers

■ **Example**. We want to find the maximum and minimum values of the function $f(x, y) = (x-2)y + y^2$, subject to the constraint $x^2 + y^2 - 1 = 0$. The Method of Lagrange Multipliers states that we need to solve the system of equations:

$$f_x = \lambda g_x, \quad f_y = \lambda g_y \text{ and } g = 0,$$

where g is the "constraint" function $g(x, y) = x^2 + y^2 - 1$.

```
f := (x,y) -> (x-2)*y + y^2;
g := (x,y) ->   x^2 + y^2 - 1;

solutions := fsolve(
   { diff(f(x,y),x) = p* diff(g(x,y),x),
     diff(f(x,y),y) = p* diff(g(x,y),y), g(x,y) = 0},
     {p, x, y}, maxsols=5):
```

`evalf(solutions, 4);`   # Shows only four digits of the answers.

{$p = -.6484, y = .7919, x = -.6107$},
{$p = 2.143, y = -.9739, x = -.2272$}

(Note: We use the variable **p** in this computation to stand for the multiplier λ.) We can find the values of f at each of these points with:

`subs( solutions[1], f(x,y));`   # Finds f at the first point.

   –1.440268753

`subs( solutions[2], f(x,y));`   # Finds f at the second point.

   3.117334694

So f has the minimum value –1.44027 at the first point (–0.6107, 0.7919) and the maximum value 3.11733 at the last point (–0.2272, –0.9739).

You can also see this result geometrically by drawing the contour picture of f and the constrained circle $g = 0$ together:

```
pict1 := contourplot(f(x,y), x=-1.5..1.5,
                y=-1.5..1.5, contours = 20,
                filled=true, coloring=[white,blue]):
with(plots):
pict2 := implicitplot( g(x,y) = 0,
                x=-2..2, y=-2..2, thickness = 3):
display({pict1, pict2}, scaling=constrained);
```

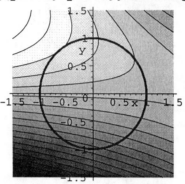

In the graphic above we used lighter shades to represent lower values of f. Thus the point $(-0.61067, 0.79189)$ on the constraint circle gives the minimum because that is where the lightest shading occurs.

Integration in Polar Coordinates

■ **Example.** Let us compute the integral $\iint_D e^{(x^2+y^2)}dA$, where D is the circular sector given in polar coordinates as $0 \le r \le 1$, $\pi/4 \le \theta \le \pi/2$. ($D$ is sketched to the right.)

We learned from calculus that we must first write the integral as the iterated integral $\int_{\pi/4}^{\pi/2}\int_0^1 e^{r^2} r\, dr\, d\theta$. Then we can compute its value:

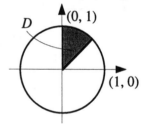

```
int(int(exp(r^2)*r, r=0..1), t=Pi/4..Pi/2);
```

$$\frac{1}{8}\, \mathbf{e}\,\pi - \frac{1}{8}\pi$$

Outsmarting Maple

■ **Example.** You can outsmart *Maple* easily. Let D be the region inside the circle $x^2 + y^2 = 1$. Find the double integral $\iint_D \sqrt[3]{x^2 + y^2}\, dA$ both by hand and using the computer.

- For the computer: Evaluate the double integral using rectangular coordinates. The region D is described by the inequalities $-1 \le x \le 1$ and $-\sqrt{1-x^2} \le y \le \sqrt{1-x^2}$, so you enter:

```
int(int((x^2+y^2)^(1/3),
        y = -sqrt(1-x^2)..sqrt(1-x^2)), x=-1..1);
```

- While you're waiting for the computer to finish, do the integration in a more "civilized" way. You can evaluate the integral directly using polar coordinates, because the region D is described by the inequalities $0 \le r \le 1$ and

$0 \le \theta \le 2\pi$:

$$\iint_D \sqrt[3]{x^2 + y^2} \, dA = \int_0^{2\pi} \int_0^1 (r^2)^{1/3} r \, dr \, d\theta = \int_0^{2\pi} \int_0^1 (r^{5/3}) \, dr \, d\theta$$

$$= 2\pi \left(\tfrac{3}{8} r^{8/3} \Big|_0^1 \right) = \tfrac{3\pi}{4}.$$

Interrupt the computer's calculation if it's still running. (If the computer has "finished," figure out whether the answer it gave was sensible!)

You Win!!!

Useful Tips

☝ ☝ If your double or triple integral is complicated, it is very easy to mistype the integrand or forget to match parentheses correctly. We recommend that you evaluate the inner and outer integrals one at a time rather than using **int** commands two or three times in a single command. You can then see which part of the integration is causing the trouble.

Troubleshooting Q & A

Question... I entered a double (or triple) integral expression and have been waiting for a response for two or three minutes. *Maple* hasn't given me an answer yet. Is something wrong?

Answer... Integration is a difficult mathematical problem. Unfortunately, *Maple* cannot solve every integration problem you encounter.

You may have to abort your calculation and look for ways to simplify the integrand (e.g., through use of cylindrical or spherical coordinates in three variables) in order to move toward getting a result.

A slow computer often takes several minutes to evaluate a complicated double or triple integral. You may simply need either a faster computer or more patience.

Question... *Maple* did not give me a number when I evaluated a double or triple integral, but I know the answer should be numerical. What should I look for?

Answer... Check for these problem areas:

- It could be a mathematical error. Did you set up the integral correctly? Recheck your limits of integration, too.

- It could be a *Maple* input error. Did you type the integrand correctly? Did you input the order of integration correctly?

- The function may have variables that are not involved in the integration and that don't evaluate as numbers. For example, your input function might be written in terms of x and y, but you're doing the integration with respect to polar coordinates **theta** and **r**.

CHAPTER 18

Matrices and Vectors

Vectors

Defining Vectors

You use lists to define vectors in *Maple*. (Please refer to Chapter 6 for elementary properties of lists.) For example, the vectors (2, 3) and (–1, 1, 2) are entered as

```
[2, 3];
```
$$[2, 3]$$

```
[-1, 1, 2];
```
$$[-1, 1, 2]$$

You do addition, subtraction, and scalar multiplication of vectors directly using the operations +, – and *. *Maple* applies these quite naturally in Release 4 or higher. However, for Release 3 or lower, you have to type the command **evalm** for the operation.

```
[1, 1, 2] + [ 2, -1, 3];          # R4 or higher.

evalm([1, 1, 2] + [ 2, -1, 3]);   # R3 or lower.
```
$$[3, 0, 5]$$

```
[1, 2, 3] - 3*[2,1,1];            # R4 or higher.

evalm([1, 2, 3] - 3*[2,1,1]);     # R3 or lower.
```
$$[-5, -1, 0]$$

Dot Product and Cross Product

You compute the dot product of two vectors using the **dotprod** command, which is defined in the **linalg** library.

```
with(linalg):
dotprod( [a1, b1, c1], [a2, b2, c2]);
```
$$a1\ a2 + b1\ b2 + c1\ c2$$

To compute a cross product of two 3-D vectors (also known as the vector product), you use the **crossprod** command, which is also defined in the **linalg** library:

```
with(linalg):
crossprod( [a1, b1, c1], [a2, b2, c2]);
```
$$[b1\ c2 - c1\ b2,\ c1\ a2 - a1\ c2,\ a1\ b2 - b1\ a2]$$

Matrices

**Defining
Matrices**

You define a matrix in *Maple* by using the **matrix** command. For example, to enter
the matrix $\begin{pmatrix} 3 & -4 & 7 \\ -1 & 0 & 5 \end{pmatrix}$, you type:

```
matrix([ [3,-4,7], [-1,0,5] ]);
```

$$\begin{bmatrix} 3 & -4 & 7 \\ -1 & 0 & 5 \end{bmatrix}$$

Notice that each row of this matrix is entered as a list of three elements. This matrix
is formed as a list of its two rows.

In general, a matrix is entered in the form:

matrix([*list of row1 entries* ,   *list of row2 entries* ,   *etc.* **]);**

> **Note:** Don't forget to use both (*parentheses*) and [*square brackets*]
> at the outer level of the matrix command. Each row must be entered
> with [*square brackets*] too.

**Basic
Operations
Involving
Matrices**

You can do addition, subtraction, scalar multiplication, and matrix multiplication
with the operators +, –, *, and **&***, respectively, together with the **evalm** command.
(The letter **m** in the name **evalm** stands for matrix.) Here are a few examples.

Operation	Example	Maple *Command*

+
Addition
	$\begin{pmatrix} 3 & -4 & 7 \\ -1 & 0 & 5 \end{pmatrix} + \begin{pmatrix} 1 & 2 & 3 \\ 4 & 5 & 6 \end{pmatrix}$	`A := matrix([[3,-4,7],[-1,0,5]]):`
		`B := matrix([[1,2,3],[4,5,6]]):`
		`evalm(A+B);`
		
		$\begin{bmatrix} 4 & -2 & 10 \\ 3 & 5 & 11 \end{bmatrix}$

–
Subtraction
	$\begin{pmatrix} 3 & -4 \\ -1 & 0 \\ 7 & 5 \end{pmatrix} - \begin{pmatrix} 1 & 2 \\ 3 & 4 \\ 5 & 6 \end{pmatrix}$	`A := matrix([[3,-4],`
		`       [-1,0],[7,5]]):`
		`B:= matrix([[1,2],[3,4],[5,6]]):`
		`evalm(A-B);`
		
		$\begin{bmatrix} 2 & -6 \\ -4 & -4 \\ 2 & -1 \end{bmatrix}$

*****
Scalar
multiplication
	$5 \begin{pmatrix} 3 & -4 & 7 \\ -1 & 0 & 5 \end{pmatrix}$	`evalm(5 * matrix([[3,-4,7],`
		`       [-1,0,5]]));`
		
		$\begin{bmatrix} 15 & -20 & 35 \\ -5 & 0 & 25 \end{bmatrix}$

&*
Matrix
multiplication
	$\begin{pmatrix} 3 & -4 & 7 \\ -1 & 0 & 5 \end{pmatrix} \cdot \begin{pmatrix} 2 & 0 & 1 \\ -1 & 1 & 2 \\ 3 & 5 & -2 \end{pmatrix}$	`A := matrix([[3,-4,7],[-1,0,5]]):`
		`B := matrix([[2,0,1],[-1,1,2],`
		`       [3,5,-2]]):`
		`evalm(A &* B);`
		
		$\begin{bmatrix} 31 & 31 & -19 \\ 13 & 25 & -11 \end{bmatrix}$

^
 	
 Power of a
 matrix
	$$\begin{pmatrix} 1 & 2 \\ 3 & 4 \end{pmatrix}^5$$	```A := matrix([[1,2],[3,4]]) :```
		 ```evalm(A^5);```
		
		 $$\begin{bmatrix} 1069 & 1558 \\ 2337 & 3406 \end{bmatrix}$$

Some Useful Matrix Commands

Maple has several commands that let you easily find the inverse, the determinant, the eigenvalues, and the eigenvectors of a given square matrix. All of these commands are defined in the **linalg** library.

Here's a quick summary:

      with(linalg):   # Needed if you have not loaded the package yet.

Operation	Maple *Command and Example*

inverse
 	
 – Find the inverse of a matrix.
	```A := matrix([[2,5,1], [3,1,2], [-2,1,0]]):```
	 ```inverse(A);```
	
	 $$\begin{bmatrix} \frac{2}{19} & \frac{-1}{19} & \frac{-9}{19} \\ \frac{4}{19} & \frac{-2}{19} & \frac{1}{19} \\ \frac{-5}{19} & \frac{12}{19} & \frac{13}{19} \end{bmatrix}$$

det
 	
 – Find the determinant of a matrix.
	```B := matrix([[2,5,1], [3,1,2], [-2,1,0]]):```
	 ```det(B);```
	
	 -19

eigenvals
 	
 – Find the eigenvalues of a matrix.
	```C := matrix([[1,2,-1], [2,3,1], [1,0,2]]):```
	 ```eigenvals(C);```
	
	 $$1, \quad \frac{5}{2} + \frac{1}{2}\sqrt{13}, \quad \frac{5}{2} - \frac{1}{2}\sqrt{13}$$

eigenvects
 	
 – Find the eigenvectors of a matrix.
	```C := matrix([[2,1,0], [-1,0,1], [1,3,1]]):```
	 ```eigenvects(C);```
	
	 $[2, 2, \{[1, 0, 1]\}], [-1, 1, \{[1, -3, 4]\}]$
	
	 This means that the eigenvalue 2 has multiplicity 2 with eigenvector (1, 0, 1); eigenvalue –1 has multiplicity 1 with eigenvector (1, –3, 4).

Elementary Row Transformations

Row Vectors

You can easily retrieve the row vector of a matrix using the **row** command defined in the **linalg** library.

      A := matrix([[2, 5, 1], [3, 1, 2],[-2, 1, 0]]);

$$\begin{bmatrix} 2 & 5 & 1 \\ 3 & 1 & 2 \\ -2 & 1 & 0 \end{bmatrix}$$

```
with(linalg):
row(A, 1);                          # This gives the first row of the matrix.
```
$$[2, 5, 1]$$

```
row(A, 3);                          # This gives the third row of the matrix.
```
$$[-2, 1, 0]$$

Row Operations

Recall from linear algebra that you often apply row transformations to a matrix to change its form into another, simpler form. *Maple* can perform row transformations using the **swaprow**, **mulrow** and **addrow** commands in the **linalg** library, as shown below:

Row Transformation	Maple *Command*
Interchange row i with row j in matrix A.	`with(linalg):` `swaprow( A, i, j );`
Multiply row i by m in matrix A.	`mulrow( A, i, m );`
Replace row j by "m *row i + row j" in matrix A.	`addrow( A, i, j, m );`

■ **Example.** Consider the matrix $\begin{pmatrix} 2 & 1 & -1 & 1 \\ 1 & 0 & 3 & 4 \\ -5 & -3 & 1 & 2 \end{pmatrix}$.

```
with(linalg):
matrix([[2,1,-1,1], [1,0,3,4], [-5,-3,1,2]]) ;
```
$$\begin{bmatrix} 2 & 1 & -1 & 1 \\ 1 & 0 & 3 & 4 \\ -5 & -3 & 1 & 2 \end{bmatrix}$$

Step 1: We would like to interchange Row 1 and Row 2.

```
swaprow( ", 1, 2);                  # Use % instead of " in R5.
```
$$\begin{bmatrix} 1 & 0 & 3 & 4 \\ 2 & 1 & -1 & 1 \\ -5 & -3 & 1 & 2 \end{bmatrix}$$

Step 2: We would like to replace Row 2 by "–2*Row 1 + Row 2."

```
addrow( ", 1, 2, -2);               # Use % instead of " in R5.
```
$$\begin{bmatrix} 1 & 0 & 3 & 4 \\ 0 & 1 & -7 & -7 \\ -5 & -3 & 1 & 2 \end{bmatrix}$$

Step 3: We would like to replace Row 3 by "5*Row 1 + Row 3."

```
addrow( ", 1, 3, 5 );               # Use % instead of " in R5.
```

$$\begin{bmatrix} 1 & 0 & 3 & 4 \\ 0 & 1 & -7 & -7 \\ 0 & -3 & 16 & 22 \end{bmatrix}$$

Step 4: We would like to replace Row 3 by "3*Row 2 + Row 3."

addrow(", 2 ,3, 3);   # Use % instead of " in R5.

$$\begin{bmatrix} 1 & 0 & 3 & 4 \\ 0 & 1 & -7 & -7 \\ 0 & 0 & -5 & 1 \end{bmatrix}$$

Step 5: We would like to replace Row 3 by "–1/5 * Row 3."

mulrow(", 3, -1/5);   # Use % instead of " in R5.

$$\begin{bmatrix} 1 & 0 & 3 & 4 \\ 0 & 1 & -7 & -7 \\ 0 & 0 & 1 & \frac{-1}{5} \end{bmatrix}$$

Bingo!! We have transformed the original matrix into a nicer form. This is called the **row echelon form** of a matrix in linear algebra.

Useful Tips

 The **linalg** library defines many other useful commands. It contains commands such as **nullspace, orthog, innerprod, jordan, kernel, minor, adjoint, BlockDiagonal, GramSchmidt, JordanBlock, LUdecomp** and **QRdecomp**. These are helpful in studying matrix theory or linear algebra.

Troubleshooting Q & A

Question... When I tried to define a matrix, I got the error message "1st and 2nd arguments (dimensions) must be non negative integers." What went wrong?

Answer... Make sure that you entered the format of a matrix correctly. Note that each row is entered as a list with [*square brackets*], all the rows are listed inside a pair of [*square brackets*], and the entire matrix is enclosed in (*parentheses*) for the **matrix** command.

Question... I tried to add, subtract, or multiply two matrices, but *Maple* returned my input unevaluated. What should I check for?

Answer... Here are some suggestions:

- First, check that you entered the matrix correctly.

- Second, you have to use the command **evalm** in order to see the result of the operation.

Question...   When I used one of the **det, eigenvals, eigenvects,** or **inverse** commands, *Maple* returned the input unevaluated. What should I check for?

Answer...   Most likely you forgot to load the **linalg** library before you used any of these commands. Type:

```
with(linalg);
```

Also, these commands only work for square matrices (same number of rows and columns), so check that too. Usually, in such circumstances, you will get the error message "square matrix expected."

Question...   When I loaded the **linalg** package

```
with(linalg);
```

I got the warning messages, "new definition for norm, ... new definition for trace." What does that mean?

Answer...   The built-in **norm** command is used to find the norm of a polynomial, and the **trace** command is used in programming. However, there are two commands in the **linalg** package also named **norm** and **trace** for linear algebra and matrix computation. Once the **linalg** library is loaded, these two new definitions will automatically replace the original definitions. You can see their difference with:

```
?norm
?linalg[norm]
?trace
?linalg[trace]
```

Question...   I entered a row transformation incorrectly, but when I retyped the correct command, *Maple* still did not give the right answer. Why is that?

Answer...   Each execution of a row transformation is done "in place," changing the matrix right away. For example, if you want to replace Row 3 by "2*Row 1 + Row 3," but instead you type (in Release 4):

```
addrow( ", 1, 3, 5) ;   # replace Row 3 by "5*Row 1 + Row 3"
```

The damage is already done. Row 3 has now been altered, and the matrix has been changed. Even if you now type the correct expression:

```
addrow( ", 1, 3, 2) ;
```

this only makes the situation worse! You need to reenter the matrix and all the previous row transformations from the beginning. (Fortunately, this is very easy to do in the Worksheet interface. Just go back, reenter the original matrix, reexecute the commands, and do the correct row transformations.)

CHAPTER 19

Parametric Curves and Surfaces in Space

Parametric Curves in Space

The spacecurve Command

You use the **spacecurve** command defined in the **plots** library to draw a parametric curve in space. To see the curve given as $(x(t), y(t), z(t))$, for $a \leq t \leq b$, type:

```
with(plots):
spacecurve( [x(t),  y(t),  z(t), t = a..b] );
```

This format is similar to the syntax of the **plot** command used for plane parametric curves. Here, however, the curve is defined with *three* parametric functions rather than *two*.

■ **Example.** The helix given parametrically by $(t, 3\cos(t), 3\sin(t))$, for $0 \leq t \leq 8\pi$, is drawn with:

```
with(plots):
spacecurve( [t, 3*cos(t), 3*sin(t), t = 0..8*Pi] );
```

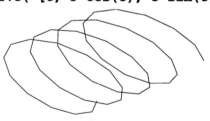

Options for spacecurve

Most options you can use with **plot3d** such as **axes, labels, color**, and **scaling** also work with **spacecurve**. By specifying a value for the **numpoints** option, you can control the resolution of the picture you see. Here's a nicer picture of the same helix.

```
spacecurve( [t, 3*cos(t), 3*sin(t), t = 0..8*Pi],
        axes = normal,
        scaling = constrained,
        color = black,  thickness= 3,
        numpoints = 100);
```

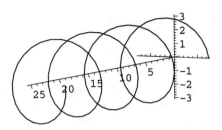

Parametric Surfaces in Space

The plot3d Command

The **plot3d** command that we used to draw the graph of a two-variable function in Chapter 15 can also be used to draw a parametric surface in space. If a surface is defined parametrically by $(x(u,v),\ y(u,v),\ z(u,v))$ for $u_0 \le u \le u_1$ and $v_0 \le v \le v_1$, you enter:

```
plot3d([{ x(u,v),  y(u,v),  z(u,v)], u = u₀..u₁, v = v₀..v₁);
```

■ **Example.** To see a portion of the one-sheeted hyperboloid given parametrically by $(\cos(u)\cosh(v),\ \sin(u)\cosh(v),\ \sinh(v))$, for $0 \le u \le 2\pi$ and $-2 \le v \le 2$, you write:

```
plot3d( [cos(u)*cosh(v),sin(u)*cosh(v),sinh(v)],
          u = 0..2*Pi, v = -2..2);
```

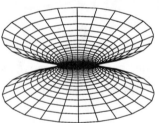

The surface $\left(v(2-\cos(4u))\cos(u),\ v(2-\cos(4u))\sin(u),\ v^2\right)$, for $0 \le u \le 2\pi$ and $0 \le v \le 2$, gives a very nice picture of a vase:

```
plot3d(
  [v*(2-cos(4*u))*cos(u),v*(2-cos(4*u))*sin(u),v^2,
        u = 0..2*Pi, v = 0..2,
        grid = [60, 30]);
```

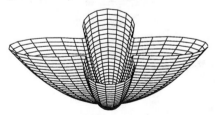

Plotting Multiple Curves and Surfaces

The **spacecurve** (or **plot3d**) command can sketch several curves (or surfaces) with a single command, just as you've already seen in the **plot**, **polarplot**, and **implicitplot** commands. Use one of these formats:

```
spacecurve({ [ curve1, t = a_1..b_1], [curve2 , t = a_2..b_2]});
```

or

```
plot3d({ [surface1 ], [surface2 ]}, u = u_0..u_1, v = v_0..v_1);
```

Multiple Curves

■ **Example.** Consider the helixes $(t, -3\sin(t), 3\cos(t))$ with $\pi \le t \le 6\pi$ and $(t, 3\cos(t), 3\sin(t))$ with $0 \le t \le 8\pi$. They can be drawn together using:

```
spacecurve( { [t,-3*sin(t),3*cos(t), t=Pi..6*Pi],
               [t,3*cos(t),3*sin(t), t=0..8*Pi] },
               numpoints = 100, color = black);
```

Multiple Surfaces

■ **Example.** The paraboloid $(r\cos t, r\sin t, 2 - r^2)$ opens down, while the paraboloid $(r\cos t, r\sin t, r^2)$ opens up. We can combine them to form a nice "beehive."

```
plot3d( { [r*cos(t),r*sin(t),r^2],
           [r*cos(t),r*sin(t), 2-r^2] },
           r = 0..1, t = 0..2*Pi);
```

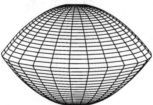

Shading and Coloring

Shading for Surfaces

The **plot3d** command illuminates a surface according to a default lighting scheme. This causes different portions of the surface to be colored or shaded differently.

You can turn off *Maple*'s default shading by setting **shading = none**. This will give an "opaque, plain" image. (Most 3-D pictures in this book were shown this way to present a clearer figure.) For example:

```
plot3d(  [2*cos(t),2*sin(t),z], t = 0..2*Pi, z=-4..4,
         scaling = constrained, shading=none);
```

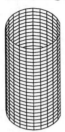

You can also control the shading more directly by specifying, for example, **shading = zgreyscale**. This will shade the cylinder above according to its height, with darker shading at the bottom:

```
plot3d(  [2*cos(t),2*sin(t),z], t = 0..2*Pi, z=-4..4,
        scaling = constrained, shading=zgreyscale,
        style = patch);                    # For release 4 or higher.
```

Coloring Function

You can explicitly set the coloring of points on a surface by using the **color** option and the HUE coloring scheme. For example,

```
s := (u,v) -> [u^2, v, v^3];
plot3d( s(u,v), u=-2..2, v=-2..2,
   style = patch, color = [(u+v)/4, u^2/4, sin(v)^2]);
```

This means that at each point $s(u,v)$, it is painted with the color defined by $\text{HUE}\left[\frac{u+v}{4}, \frac{u^2}{4}, \sin^2(v)\right]$. For example, at $s(0,1)$ the color is $\text{HUE}[1/4, 0, \sin^2 1]$.

More Examples

Combining Graphics with Show

You can combine curves, surfaces, and any other three-dimensional images into a single graphic using the **display** command (just as we saw in Chapter 15).

■ **Example.** The upper hemisphere of the unit sphere $x^2 + y^2 + z^2 = 1$ is given by $(\, r\cos(t), r\sin(t), \sqrt{1-r^2}\,)$, for $0 \le r \le 1$ and $0 \le t \le 2\pi$.

```
g1 := plot3d( [r*cos(t), r*sin(t), sqrt(1-r^2)],
            r = 0..1, t = 0..2*Pi):
```

The point $P = (\frac{1}{2}, \frac{1}{2}, \frac{1}{\sqrt{2}})$ lies on this hemisphere. A normal (perpendicular) line to the hemisphere at P is given by $\vec{r}(t) = (\frac{1}{2}+t, \frac{1}{2}+t, \frac{1}{\sqrt{2}}+\sqrt{2}\,t)$. This command shows just a portion of the normal line:

```
with(plots):
g2 := spacecurve( [1/2+t,1/2+t, 1/sqrt(2)+sqrt(2)*t],
        thickness = 3, t=0..0.15):
```

Finally, the plane tangent to the hemisphere at P has equation $z = \frac{2-x-y}{\sqrt{2}}$. You can sketch a portion of it near the point P with:

```
g3 := plot3d( (2-x-y)/sqrt(2),
        x = 0.2..0.8, y = 0.2..0.8, grid =[2,2]):
```

You can now see one of the nicest features of *Maple*, the ability to combine these graphics, despite the fact that each was drawn using a different type of command.

```
display({g1,g2,g3}, orientation = [97,78],
        scaling = constrained);
```

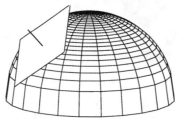

Notice how well the graphic convinces you that we have described both the tangent plane and the normal line correctly.

Troubleshooting Q & A

Question...   When I tried to use the **spacecurve** command, *Maple* returned the input unevaluated. What went wrong?

Answer...   You forgot to load the **plots** library first. Type:

```
with(plots);
```

Question...   When I tried to draw a parametric curve or surface in space, I got an error message. What should I check?

Answer...   There are many possibilities, but our best suggestions are:

- Did you use the **spacecurve** or **plot3d** command? A common mistake is trying to draw a 3-D picture with the **plot** command.

- Check that you have followed the correct syntax of entering the commands. The formats for **spacecurve** and **plot3d** are different and can cause confusion.

- Did you enter the function(s) correctly without a typing mistake? Is each of the functions defined everywhere in the interval(s) you specified?

- Did you use the same literal parameter in both your function and interval? (For example, check that you didn't write **f(x,y)** when the parameters were **u** and **v**.)

- Did you use two parameters for a surface?

Question... When I used **plot3d** to draw *one* parametric surface, I got *three* surfaces instead. What happened?

Answer... On some rare occasions, it can happen that your parametric surface is actually split into three pieces. However, most likely, you entered the command incorrectly. A common mistake is to type { *curly braces* } instead of [*square brackets*] for the coordinate functions. *Maple* will then interpret your input as a request to draw three *graphs* instead (see Chapter 15).

CHAPTER 20

Vector Fields

Drawing a Vector Field

The fieldplot Command

The **fieldplot** command (defined in the **plots** library) sketches vector fields in the plane. To see the vector field defined by $\vec{F}(x, y) = (F_1(x, y), F_2(x, y))$ for $x_0 \leq x \leq x_1$, $y_0 \leq y \leq y_1$, you enter:

```
with(plots);
fieldplot( [ F₁(x, y), F₂(x, y) ], x = x₀..x₁, y = y₀.. y₁);
```

For example, to see the vector field $(-y, x)$, for $-2 \leq x \leq 2$, $-2 \leq y \leq 2$, you type:

```
with(plots);
fieldplot([-y,x], x=-2..2, y=-2..2,
                   scaling=constrained);
```

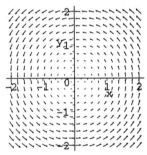

Strictly speaking, *Maple* does not draw the vector field $(-y, x)$ exactly. It has scaled the lengths of the vectors proportionally to produce a nice picture. Also, *Maple* draws a heavier arrow when the vector is longer.

Some Useful Options

Some options for the **fieldplot** command can help us change the style of the output picture.

Option	What It Does
`grid = [10,10]`	Draws $10 \times 10 = 100$ vectors. Default value is **[20, 20]**.
`arrows = thick`	Uses thicker arrows for vectors.
`color = blue`	The arrows will be drawn in blue (or whatever color you specify).

For example, here's a different picture of the vector field above:

```
fieldplot([-y,x], x=-2..2, y=-2..2, grid = [10,10],
                     arrows = thick, color = grey);
```

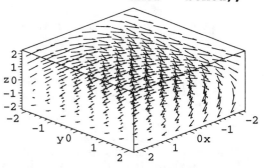

Vector Fields in Space

To draw a 3-D vector field (in space), you use the **fieldplot3d** command. It is defined in the **plots** library.

To see the vector field $\vec{F}(x, y, z) = (F_1(x, y, z),\ F_2(x, y, z),\ F_3(x, y, z))$, for the intervals $x_0 \le x \le x_1$, $y_0 \le y \le y_1$, and $z_0 \le z \le z_1$, you type:

```
with(plots):
fieldplot3d( [F_1(x, y, z),  F_2(x, y, z),  F_3(x, y, z)],
             x = x_0 .. x_1,  y = y_0 .. y_1,  z = z_0 .. z_1);
```

For example, to see the vector field $(-z,\ 1,\ x)$, use:

```
with(plots):
fieldplot3d( [-z,1,x], x=-2..2, y=-2..2, z=-2..2,
                       axes = boxed);
```

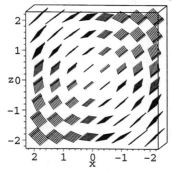

This looks like a mess. You can get a better picture by looking from the positive, y-axis. It looks like the vector field is swirling around the y-axis.

Gradient, Curl, and Divergence

Gradient Field   The gradient of a function is denoted by grad f. It is defined to be the vector

$$\text{grad}\, f = (f_x, f_y), \text{ or grad}\, f = (f_x, f_y, f_z),$$

depending on whether f is a function of two or three variables, respectively. You can find the gradient of f by calculating the partial derivatives explicitly:

```
f := (x,y,z) -> x^2 + 2*x*y - z^3;
[ diff(f(x,y,z),x), diff(f(x,y,z), y),
                    diff(f(x,y,z),z)] ;
```

$$[2x+2y, 2x, -3z^2]$$

You can also use the **grad** command from the **linalg** library. This lets you find the gradient in any standard coordinate system (e.g., cartesian (rectangular), spherical, or cylindrical, and some esoteric ones too!).

The following table gives examples of computing the gradient in the three most-used coordinate systems. But first, we load the package that defines the **grad** command:

```
with(linalg):
```

Coordinate System	Maple *Command*
Cartesian or rectangular(x, y, z)	`grad( x^2 + 2*x*y - z^3, [x,y,z],` `         coords = cartesian);` # R3, R4 `grad( x^2 + 2*x*y - z^3, [x,y,z],` `         coords = rectangular);` # R5 $[2x+2y,\; 2x,\; -3z^2]$

Cylindrical (r, θ, z)	`grad(r*cos(theta)+r^2*z, [r,theta,z],`
	`         coords = cylindrical);`
	
	$[\cos(\theta) + 2rz,\; -\sin(\theta),\; r^2]$

Spherical (ρ, θ, ϕ)	`grad(2*rho^2*cos(theta)*sin(phi),`
	`  [rho, theta, phi], coords = spherical);`
	
	$\left[4\rho \cos(\theta) \sin(\phi)\, ,\, -2\rho\sin(\theta)\sin(\phi)\, ,\, 2\frac{\rho\cos(\theta)\cos(\phi)}{\sin(\theta)} \right]$

> **Note:** In the commands above, the order of the variables such as **[x, y, z]**, **[r, theta, z]**, and **[rho, theta, phi]** is important. If you list them in a different order, the computation of the gradient will be incorrect.

Curl and Divergence   To calculate the curl and divergence of a vector field, use the **curl** and **diverge** commands, respectively. They are also defined in the **linalg** package and are used much as the **grad** command. First load them with:

```
with(linalg):
```

Examples	Maple *Command*
Divergence in cartesian (rectangular) system	`diverge([x^2,y^2,z^2], [x,y,z],` `         coords = cartesian);` # for R3 or R4 $2x + 2y + 2z$

Curl in cartesian (rectangular) system	`curl( [x+y, y+z, sin(x*y)+z^2], [x,y,z],`
	`      coords = rectangular);   # for R5`
	
	$\left[\cos(x\,y)\,x - 1, \ -\cos(x\,y)\,y, \ -1\right]$

Divergence in spherical system	`diverge([rho^2, rho*sin(phi), sin(theta)],`
	`        [rho,phi,theta], coords = spherical);`
	
	$\dfrac{4\,\rho^2\,\sin(\phi) + 2\,\rho\,\sin(\phi)\cos(\phi) + \cos(\theta)}{\rho\,\sin(\phi)}$

Line and Surface Integrals

Integration of a Vector Field

Engineers are often interested in integrating a vector field $\vec{F}$ either along a curve $\vec{r}(t)$, for $a \le t \le b$, or over a surface $\vec{s}(u,v)$, for $u_0 \le u \le u_1$, $v_0 \le v \le v_1$. These are defined as:

- Line integral: $\int_a^b \vec{F}(\vec{r}(t)) \cdot \vec{r}'(t) \, dt$

- Surface integral: $\pm \int_{u_0}^{u_1} \int_{v_0}^{v_1} \vec{F}(\vec{s}(u,v)) \cdot \left(\dfrac{\partial \vec{s}}{\partial u} \times \dfrac{\partial \vec{s}}{\partial v} \right) dv \, du$ (The choice of the $\pm$

  sign depends on how the normal vector for the surface is defined.)

These integrals are easily computed using the **int** command, as we show in the following examples.

Line Integral

■ **Example.** Let $\vec{F}$ be the vector field $\vec{F}(x, y) = (x + y, -y)$ and $\vec{r}(t)$ the parametric curve $\vec{r}(t) = (1 - t, t^2)$ for $0 \le t \le 1$. The line integral of $\vec{F}$ over this curve is computed as:

```
F := (x,y) -> (x+y, -y):    # [F(x,y)] defines the vector field F.
r := t -> (1-t, t^2 ):      # [r(t)] defines the curve r(t).

with(linalg):               # We will use the dotprod command.
int(dotprod([F(r(t))], diff([r(t)], t)), t=0..1);
```

$$-\frac{4}{3}$$

> **Note:** Syntax is important! We defined **F** and **r** using (*parentheses*) instead of [*square brackets*]. This lets us compute $\vec{F}(\vec{r}(t))$ easily with the expression **[F(r(t))]**, although we now also have to use **[F(x,y)]** and **[r(t)]** to represent $\vec{F}(x, y)$ and $\vec{r}(t)$, respectively.

We can see why the line integral turned out to be negative by checking how the vector field lines up with the curve.

```
with(plots):
field := fieldplot( [F(x,y)], x=0..1, y=0..1,
                         grid=[10,10]):
curve := plot( [r(t), t=0..1], thickness = 3):
display({field, curve}, scaling=constrained);
```

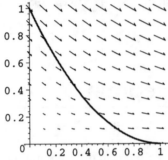

As you can see from the picture, if a particle moves along the curve from its starting point $\vec{r}(0) = (1,0)$ to its ending point $\vec{r}(1) = (0, 1)$, it is moving against the force of the vector field. Thus, the line integral is negative.

Surface Integral

■ **Example.** Suppose the surface integral of $\vec{F}(x, y, z) = (x(1+z), y, 0)$ over the surface S is given by

$$-\int_1^4 \int_0^{2\pi} \vec{F}(\vec{s}(u,v)) \cdot \left(\frac{\partial \vec{s}}{\partial u} \times \frac{\partial \vec{s}}{\partial v} \right) dv\ du$$

where S is parameterized by $\vec{s}(u,v) = (u\cos v, u\sin v, u)$. We can evaluate this integral as follows:

```
F := (x,y,z) -> (x*(1+z), y, 0):
s := (u, v) -> (u*cos(v), u*sin(v), u) :

su := diff([s(u,v)],u);      # This calculates ∂s⃗ / ∂u.
sv := diff([s(u,v)],v);      # This calculates ∂s⃗ / ∂v.

with(linalg):               # We need the crossprod command.

- int(int( dotprod( [F(s(u,v))], crossprod(su, sv)),
            v=0..2*Pi), u=1..4);
```

$$\frac{423\pi}{4}$$

More Examples

The Perpendicular Property of the Gradient

■ **Example.** Consider the function $f(x, y) = xy + 2x$:

```
f := (x,y) -> x*y +2*x;
```

The perpendicular property of the gradient vector states that grad $f(a,b)$ is perpendicular to the level curve of f that goes through (a,b). To see this, we first draw several level curves:

```
pict1 := contourplot(f(x,y), x=-4..4, y=-4..4,
               contours = 20, color = black):
```

By direct computation, grad $f(x, y) = (y + 2, x)$. We can draw this gradient field with the **fieldplot** command.

```
with(plots):
pict2 := fieldplot([y+2,x], x=-4..4, y=-4..4,
                    grid = [10,10]):
```

Now combine the contours and gradient vectors into one picture:

```
display({pict1, pict2}, scaling=constrained);
```

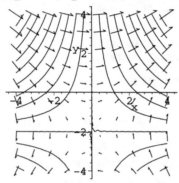

Note: It may take a while for *Maple* to show the picture above!

Do you agree that the gradient vectors are perpendicular to the level curves at every point?

Useful Tips

 In many cases, you can draw a clearer (and faster) picture of vector fields by using a smaller number of gridlines, such as **grid = [10, 10]**.

Troubleshooting Q & A

Question... I did not get a picture from **fieldplot** or **fieldplot3d**. What should I check for?

Answer... There are several possible mistakes.

Input errors:

• Did you load the **plots** library? If you forgot, type the following:

```
with(plots);
```

• Did you mistakenly use **fieldplot** for a 3-D vector field or **fieldplot3d** for a 2-D vector field?

- Check that you have input the correct format and expression. For example, a common mistake is to enter the vector field $(x, x + y)$ as **(x, x+y)** instead of **[x, x+y]**.

Hardware problems:

- Does the computer have enough memory to draw the picture (especially if you have already drawn a lot of graphics)? Restarting *Maple* may help.
- If your computer is slow, it can take a long time to show the output. Reducing the number of vectors by using the **grid** option can dramatically improve speed.

Question... When I used **grad**, **curl**, or **diverge**, *Maple* returned my input unevaluated. What was my mistake?

Answer... You forgot to load the **linalg** package before you used these commands. Correct the problem by loading the package:

```
with(linalg);
```

Question... I got a wrong answer when I used **grad**, **curl**, or **diverge**. What was my mistake?

Answer... There are several possibilities:

- Check that the input syntax is correct.
- You have to type the vectors **[x, y, z]**, or **[r, theta, z]**, or **[rho, theta, phi]** with the variables exactly in this order. If you list them in a different order, the computation of the gradient, curl, or divergence will be incorrect.
- The input function or vector field has to be expressed in terms of the variables of your chosen coordinates. For example, if you want to calculate the gradient of $x + y + z$ using cylindrical coordinates, you have to enter the function as **r*cos(theta)+r*sin(theta) +z**.

CHAPTER 21
Basic Statistics

The **stats** (statistics) library in *Maple* contains a multitude of commands and procedures to analyze and describe data. Unlike the other libraries, it is organized in "sublibraries" that can be called independently. In this chapter and the next, we will show you some basic statistics tools that you can use. Once you are familiar with the underlying concepts, the use of other advanced statistics tools is generally straightforward.

> **Note:** The statistics commands we describe in this chapter are for Release 3 or higher. These commands may work differently in Release 1 or 2.

Graphical Presentation of Data

Scatter Diagrams

A set of numerical data can sometimes be presented graphically by using a scatter diagram. Depending on the type of data, you can either use **pointplot** (defined in the **plots** library) or **scatterplot** (defined in the sublibrary **statplots** of the **stats** library) to see the diagram. (In Releases 3 and 4, **scatterplot** is called **scatter2d** instead.)

```
with(plots);
pointplot( a list of ordered pairs );

with(stats[statplots]):
scatterplot( list of x-coordinates, list of y-coordinates);      # for R5
scatter2d( list of x-coordinates, list of y-coordinates);        # for R3, R4
```

For example,

```
with(plots):
pointplot( [[1, 3], [2, 1], [6,2], [4, -1]]);
```

```
with(stats[statplots]):
scatterplot( [1,2,6,4], [3,1,2,-1]);      # R5
scatter2d( [1,2,6,4], [3,1,2,-1]);        # R3, R4
```

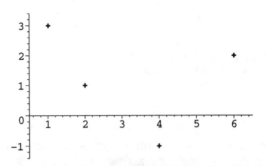

Note: You use the **pointplot** command if the data is a list of pairs. You use the **scatterplot** (or **scatter2d**) command if you have separate lists of *x*- and *y*-coordinates.

Numerical Measures of Data

Mean, Median, Standard Deviation, etc.

Standard statistical measures of data distribution are available in *Maple*. To use them, you first have to load the **describe** sublibrary of the **stats** library:

```
with(stats[describe]);
```

  [coefficientofvariation, count, countmissing, covariance, decile, geometricmean, harmonicmean, kurtosis, linearcorrelation, mean, meandeviation, median, mode, moment, percentile, quadraticmean, quantile, quartile, range, skewness, standarddeviation, sumdata, variance]

■ **Example.** Let us consider the number of home runs for the 14 baseball teams in the National League for the 1997 season.

```
homeRuns := [174, 239, 142, 127, 136, 133, 174,
             172, 153, 129, 116, 152, 172, 144];
```

We can find the mean and median directly with the following commands.

```
mean(homeRuns);
```

$$\frac{309}{2}$$

```
median(homeRuns);
```

  148

Other measures of central tendency such as the **harmonicmean, geometricmean**, and **mode** are also available and work in the same way.

The best-known measures of data variability also have their customary names:

```
evalf(variance(homeRuns));
```

  891.5357143

```
evalf(standarddeviation(homeRuns));
```

  29.85859531

```
evalf(meandeviation(homeRuns));
```

$$22.64285714$$

Maple's default on variance and standard deviation is for full populations. For example, the variance is calculated as $\frac{1}{N}\sum_{i=1}^{N}(X_i - \overline{X})^2$. If you want the sample variance $\frac{1}{N-1}\sum_{i=1}^{N}(X_i - \overline{X})^2$, you have to add an index of 1:

evalf(variance[1](homeRuns));   # Sample variance.

$$960.1153846$$

evalf(standarddeviation[1](homeRuns));

$$30.98572872$$

Probability Distributions

The standard statistics libraries contain definitions of common probability distributions. These include the following, along with their usual parameter specifications:

normald[μ, σ]	uniform[a, b]	binomiald[λ]
studentst[n]	poisson[λ]	beta[α, β]
exponential[λ]	discreteuniform[a,b]	chisquare[n]

Cumulative Distribution and Probability Density Functions

You can work with the **cumulative distribution function**, the **inverse cumulative distribution function**, and the **probability density function** of each of these distributions using the **statevalf** command (defined in the **stats** library). The syntax is:

```
with(stats);
statevalf[ function type, distribution ]( arguments );
```

where the *function type* is listed in the following table:

Type of Distribution	function type	Meaning
Continuous	cdf	Cumulative distribution function
	icdf	Inverse cumulative distribution function
	pdf	Probability density function
Discrete	dcdf	Discrete cumulative distribution function
	idcdf	Inverse discrete cumulative distribution function
	pf	Probability density function

> **Note:** You have to use [*square brackets*] to surround the function type and distribution name in the **statevalf** command, but (*parentheses*) for the arguments of the command.

■ **Example.** Consider the normal distribution with mean 10 and deviation 3.

```
with(stats):
distribution1 := normald[10,3];
```
$$distribution1 := normald_{10,\,3}$$

To find the values of its cumulative distribution function and probability density function, say at $x = 15$, use:

```
statevalf[cdf,distribution1](15);
```
.9522096478

```
statevalf[pdf,distribution1](15);
```
.03315904623

Also, we can see the graphs of the cumulative distribution and probability density functions, respectively, with:

```
plot(statevalf[cdf,distribution1](x), x=0..20);
```

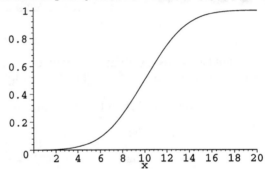

```
plot(statevalf[pdf,distribution1](x), x=0..20);
```

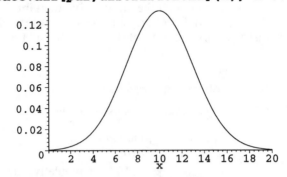

The inverse cumulative probability distribution can also be graphed: (Note that the domain of an inverse cumulative probability distribution function is $0 \le x \le 1$.)

```
plot(statevalf[icdf,distribution1](x), x=0..1);
```

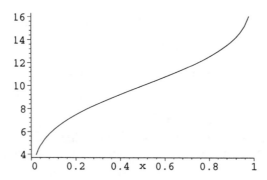

■ **Example.** We need to use **dcdf** and **pf** when working with discrete distributions such as the binomial distribution:

```
distribution2 := binomiald[15,.7]:
```

To find the value of its cumulative distribution function at 10:

```
statevalf[dcdf,distribution2](10);
```
.4845089408

We can also plot the "graph" of its probability density function:

```
pointplot([seq(
        [i, statevalf[pf,distribution2](i)],i=0..15)],
        title= `Binomial probs, n=15, p=.7`);
```

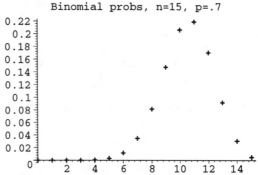

You can work with other distributions in a similar way. All you need to do is to include the proper parameter(s) when you reference the distributions.

More Examples

IQ Score

■ **Example.** It is generally believed that IQ scores are normally distributed with mean 100 and standard deviation 15. Let us theoretically estimate how many people in the world have an IQ higher than 180.

The probability that a person's IQ is between 0 and 180 is:

```
statevalf[cdf,normald[100,15]](180);
```
.9999999518

So the probability that a person's IQ is higher than 180 is:

1 - ";   # or **1 – %;** for Release 5.

.482 10⁻⁷

There are about 5 billion people in this world, so:

" * 5*10^9;   # or **% * 5*10^9;** for Release 5.

241.000000

Only about 250 people in the world have IQs higher than 180. Isn't it odd, then, how often we meet people who claim to have such high IQs?

Generating a Random Sample from a Specified Distribution

■ **Example.** Using the command **random** defined in the **random** library, you can generate a random sample from specified distributions. For example, to form a list of 50 random values from the Poisson distribution with mean 5, we use:

with(random):
list1 := [random[poisson[5]](50)];

list1 := [7.0, 6.0, 6.0, 4.0, 3.0, 8.0, 4.0, 5.0, 7.0, 2.0, 5.0, 5.0, 8.0, 6.0, 8.0, 5.0, 5.0, 10.0, 4.0, 11.0, 3.0, 8.0, 4.0, 2.0, 2.0, 5.0, 3.0, 5.0, 2.0, 4.0, 6.0, 6.0, 3.0, 6.0, 3.0, 1.0, 2.0, 5.0, 2.0, 6.0, 1.0, 2.0, 6.0, 2.0, 7.0, 11.0, 6.0, 3.0, 6.0, 3.0]

The mean of this sample is:

mean(list1);

4.880000000   # This is pretty close to the theoretical value of 5.

■ **Example.** Let us consider **exponential[3,0]**. Its mean is $\frac{1}{3}$ and its variance is $\frac{1}{9}$. Suppose we generate a sample of 1000 random observations from this distribution. We can see that the mean and variance of this random sample are very close to the theoretical values of $\frac{1}{3}$ and $\frac{1}{9}$, respectively: Exponential distribution

randomdata := [random[exponential[3,0]](1000)];
mean(randomdata);

.3396637095

variance(randomdata);

.1138093963

Importing Numerical Values from a File

You can easily import data values into *Maple* from another application (e.g., a spreadsheet or an e-mail message) and then use the statistical tools we've introduced in this chapter to study the data.

For example, suppose you have a data file named "TestScores.dat" where individual data values are written in text format, one to a line. You can read all the data values into a list named **dataValues** by using the **importdata** command as follows:

with(stats);
dataValues := [importdata(`TestScores.dat`)];

Now you can work with and analyze the data in the list **dataValues** by using commands from *Maple*'s statistics packages.

Useful Tips

 When you are working with statistical analysis, you will probably also want some descriptive statistics, the ability to evaluate a distribution, and some plotting tools. We recommend that you start by loading all the commands available in the statistics library with:

with(stats); with(describe); with(statevalf); with(statplots);

You can load each individual statistics library separately. But if you load them all at the start of your *Maple* session, you're less likely to run into the usual troubles associated with unloaded commands.

Troubleshooting Q & A

Question ... I tried to evaluate a cumulative probability function or probability density function at a point, but I got an error message such as "... no subpackage function specified" or "... wrong number (or type) of parameters." What went wrong?

Answer ... The statistics commands in *Maple* require special attention to the difference between brackets and braces. The command **statevalf** uses [*square brackets*] to enclose the function and the distribution. You use [*square brackets*] for the parameters of the distribution, but (*parentheses*) for the point at which you evaluate the distribution.

statevalf[cdf,normald[0,1]](3);

Question ... Working on a desktop machine, when I tried to use **importdata** to load a file, I got an error message, "... file I/O error." What went wrong?

Answer ... The file may not exist, or it may not have the name you gave in the command. If it does exist and has the right name, check to see that is located in the folder or directory where *Maple* is looking for it.

The easiest way to verify where *Maple* looks for files is to use the **writedata** command to create a file with a distinctive name. Then, search for that filename.

writedata(`LookForMe`,[1,2,3]);

Start by looking for the file "LookForMe" in the folder with *Maple*.

CHAPTER 22

Fitting Curves to Data

Regression

Using a Least Squares Fit

Given a list of data, you may want to find the line that "best" fits this data set. (Best fit is usually determined by the "least square method.") This process is called **linear regression**. We can use the **leastsquare** command from the **stats[fit]** library to carry out linear regression.

```
with(stats[fit]):
leastsquare[[x,y], y = a*x+b ]([ x-data, y-data ]);
```

For example,

```
with(stats[fit]);
leastsquare[[x,y], y = a*x+b ]([ [1,3,5], [2, 4,7] ]);
```

$$y = \frac{5}{4} x + \frac{7}{12}$$

This means that the line $y = \frac{5}{4}x + \frac{7}{12}$ best fits the three points (1, 2), (3, 4), (5,7).

■ **Example.** Here are data from the UNESCO 1990 *Demographic Yearbook* which show male life expectancy and female life expectancy in Argentina, Austria, Afghanistan, Algeria, Angola, Bolivia, Brazil, Belgium, Bangladesh, and Botswana, respectively.

```
mdata := [65.5, 73.3, 41, 61.6, 42.9, 51, 62.3, 70,
              56.9,52.3]:
```

```
fdata := [72.7, 79.6, 42, 63.3, 46.1, 55.4, 67.6,
              76.8, 56,59.7]:
```

Now, let us find the line that best fits the data.

```
regline :=
   leastsquare[[x,y], y=a*x +b ]([mdata, fdata]);
```
$$regline := y = 1.128850251\, x - 3.192082489$$

To see how well the line fits the data, we first draw the points using **scatterplot** (or **scatter2d** for Release 3 or 4, see Chapter 21):

```
with(plots): with(stats[statplots]):
pict1 := scatterplot(mdata,fdata):
```

To plot the line, we need the right-hand side of the expression **regline** which is given by **rhs(regline)**.

```
pict2 := plot(rhs(regline), x=40..75):
display({pict1, pict2});
```

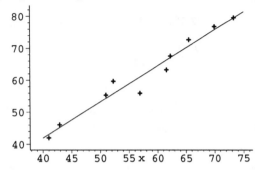

Under this model, say, if a certain country's male life expectancy is 65 years, then the female life expectancy is estimated to be:

```
subs(x=65, regline);
```

$$y = 70.18318383$$

Testing Linear Correlation

You use the command **linearcorrelation** (defined in the **describe** sublibrary of the **stats** library) to find how well the best fitting line fits the data. Using our previous example:

```
with(stats[describe]):
linearcorrelation(mdata,fdata);
```

$$.9794660676$$

This value is very close to 1. Thus, a strong correlation exists between the life expectancy of males and females in the countries.

Fitting with Other Functions

The **leastsquare** command can also work for other functions. Given functions $f_1(x), f_2(x), ..., f_n(x)$, if you want to find a function of the form

$$a_1 f_1(x) + a_2 f_2(x) + ... + a_n f_n(x)$$

that best fits a given set of data, you will use one of these formats:

> **leastsquare[[x,y]**, $y = a_1 f_1(x) + a_2 f_2(x) + ... + a_n f_n(x)$**]**
> ([*x-data*, *y-data*])**;**
>
> **leastsquare[[x,y]**, $y = a_1 f_1(x) + a_2 f_2(x) + ... + a_n f_n(x)$,
> $\{a_1, a_2, ..., a_n\}$**]** ([*x-data*, *y-data*])**;**

> **Note**: Use the first format if all the functions $f_i(x)$ are polynomials. Use the second format otherwise.

For example, using the previous data:

```
leastsquare[[x,y],y=a+b*x+c*x^2]
          ([mdata,fdata]);      # Best quadratic fit.
```

$$y = 5.744726622 + .8029638020\ x + .002871857378\ x^2$$

```
leastsquare[[x,y],y=a+b*x+c*x^2+d*x^3]
            ([mdata,fdata]);     # Best cubic fit.
```

$$y = -122.1805302 + 7.812898348\,x - .1222953133\,x^2 + .0007297988519\,x^3$$

```
leastsquare[[x,y],y=a+b*exp(2*x)+c*exp(x),{a,b,c}]
            ([mdata,fdata]);
```

$$y = 57.79885201 - .1111944598\ 10^{-60}\,e^{2x} + .7903142839\ 10^{-29}\,e^{x}$$

The last answer suggests that in this case it is not a good idea to use exponential functions to fit the given data.

Interpolation

The interp Command

Given n points (x_1, y_1), (x_2, y_2), ..., (x_n, y_n), you can find a polynomial of degree $n-1$ that perfectly matches all these values. This can be done in *Maple* using the command **interp** in the syntax:

```
interp( x-data, y-data, x);
```

For example, to find a polynomial $g(x)$ that passes through all of the points, use:

```
xdata := [0, 1, 2, 3, 4, 5, 6]:
ydata := [0, 16, -10, 28, -30, -12, -12]:
g := interp(xdata, ydata,x);
```

$$g := -\frac{299}{180}\,x^6 + \frac{299}{10}\,x^5 - \frac{1819}{9}\,x^4 + \frac{1897}{3}\,x^3 - \frac{162041}{180}\,x^2 + \frac{13733}{30}\,x$$

You can check that $g(x)$ passes through all the given points with:

```
points := [seq([i,evalf(subs(x=i,g))], i=0..6)];
```

$points := [[0, 0], [1, 16.], [2, -10.], [3, 28.], [4, -30.], [5, -12.], [6, -12.]]$

You can see that the graph of $g(x)$ passes through all the given points with:

```
with(plots):
pict1 := pointplot(points, symbol=cross):
pict2 := plot(interp(xdata, ydata,x),x=0..6):
display([pict1,pict2]);
```

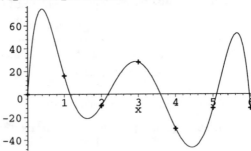

More Examples

Using a Least Median Squares Fit

A drawback with the least squares method of curve fitting is that it is extremely sensitive to a single bad data point. For fitting linear equations, an alternative is to use **leastmediansquare** to fit a curve to the data. The use of the command is similar to how we used the **leastsquare** command.

```
with(stats[fit]):
leastmediansquare[[x,y]]([ x-data, y-data ]);
```

■ **Example.** Let us revisit the UNESCO data from our earlier example but add in an extra point at (5, 85). We will then compare the results with least square and least median square fitting.

> `mdata2 := [seq(mdata[i],i=1..10), 5];`   # With extra data 5.

$mdata2 := [65.5, 73.3, 41, 61.6, 42.9, 51, 62.3, 70, 56.9, 52.3, 5]$

> `fdata2 := [seq(fdata[i],i=1..10), 85];`   # With extra data 85.

$fdata2 := [72.7, 79.6, 42, 63.3, 46.1, 55.4, 67.6, 76.8, 56, 59.7, 85]$

> `lstsq1 := leastsquare[[x,y],`
> `        y=a*x+b]([mdata,fdata]);`   # The original data.

$lstsq1 := y = 1.128850251\, x - 3.192082489$

> `lstsq2 := leastsquare[[x,y],`
> `        y=a*x+b]([mdata2,fdata2]);`   # With the bad point.

$lstsq2 := y = .02684747952\, x + 62.59819422$

Notice how different these results are when using least squares. The first is the line $y \approx 1.13x - 3.19$, while the second is $y \approx 0.03x + 62.60$. One bad point has affected the result dramatically.

However, if we use **leastmediansquare**,

> `leastmedsq1 :=leastmediansquare[`
> `        [x,y]]([mdata,fdata]);`   # The original data.

$leastmedsq1 := y = -2.042105263 + 1.126315789\, x$

> `leastmedsq2 := leastmediansquare[`
> `        [x,y]] ([mdata2,fdata2]);`   # With the bad point.

$leastmedsq2 := y = -2.042105263 + 1.126315789\, x$

Adding one point to the original data has had virtually no effect on the line computed using the least median square method.

Fitting vs. Interpolating-Polynomial

You may wonder about the difference between the methods of least squares fitting (**leastsquare**) and interpolation (**interp**). A least squares fit allows us to find a function of a specified form that best *approximates* the given set of data. On the other hand, interpolation will find a function that *matches the data exactly*.

■ **Example.** Consider the UNESCO data of the first example of this chapter. If we had used interpolation rather than finding a least squares fit for this data, we would have found the polynomial:

```
Digits := 12;      # We need to increase the number of digits used in
                   # the floating point computation, because the answer
                   # behaves so wildly, as you will see.
```

```
interp(mdata,fdata,x);
```

$$-.397572075218\ 10^{-7} x^9 +.203073369106\ 10^{-4} x^8 -.458829289271\ 10^{-2} x^7$$

$$+.601845366123\ x^6 + 50.5044284563\ x^5 + 2811.63405653\ x^4 - 103837.722471 x^3$$

$$+2453065.00239\ x^2 - 33636458.2451\ x + 203962232.163$$

Let's compare graphically the results obtained by the two methods:

```
with(plots): with(stats[statplots]):

pict1 := scatterplot(mdata, fdata):
pict2 := plot(rhs(leastsquare[[x,y]]
         ([mdata,fdata],x)), x=40..75, color=green):
pict3 := plot(interp(mdata,fdata,x),
         x=40..75,y=30..80, color=grey):

display({pict1, pict2, pict3});
```

(Use **scatter2d** instead of **scatterplot** in Release 3 or 4.)

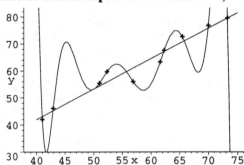

You can see in this case that the line obtained by least squares fitting is likely to be better for prediction.

A Temperature Model

■ **Example.** The following monthly data in the form [*month, temperature in* $F°$] show the average monthly high temperatures in Boston, Massachusetts (1 represents the month of January, 2 for February, and so on).

```
bostonHigh := [[1, 36.4], [2, 37.7], [3, 45.0],
               [4, 56.6], [5, 67.0], [6, 76.6],
               [7, 81.8], [8, 79.8], [9, 72.3],
               [10, 62.5],[11,51.6], [12,40.3],
               [13, 36.4],[14, 37.7],[15, 45.0]]:
pointplot(bostonHigh, symbol=cross);
```

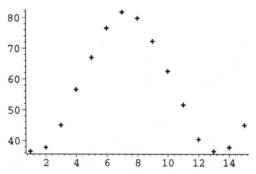

Since temperature is cyclic, we expect that it can be described nicely by a cosine function in the form

$$Temperature = a_1 + a_2 \cos(a_3(x - a_4)), \text{ where } x \text{ is the month.}$$

The temperature's period is 12 months, so we should choose $a_3 = \frac{2\pi}{12} = \frac{\pi}{6}$. Also, since we expect July to be the hottest month of the year, the function must have its largest value when $x = 7$. This means that we should choose $a_4 = 7$.

Thus, we need to find the function of the form $a_1 + a_2 \cos(\frac{\pi}{6}(x - 7))$ that best fits the temperature data. But first we rewrite the data in separate lists of x-coordinates and y-coordinates.

```
xboston := [seq(bostonHigh[i][1], i=1..15)];
```
$$xboston := [1, 2, 3, 4, 5, 6, 7, 8, 9, 10, 11, 12, 13, 14, 15]$$

```
yboston := [seq(bostonHigh[i][2], i=1..15)];
```
$$yboston := [36.4, 37.7, 45.0, 56.6, 67.0, 76.6, 81.8, 79.8,$$
$$72.3, 62.5, 51.6, 40.3, 36.4, 37.7, 45.0]$$

Now we can find the best fit function and see the picture.

```
bestfit := leastsquare[[x,y],
    y=a+b*cos(Pi/6*(x-7)), {a,b}]([xboston,yboston]);
```

$$bestfit := y = 58.7015403930 + 22.7483212182 \cos\left(\frac{1}{6}\pi(x-7)\right)$$

```
plot1 := pointplot(bostonHigh, symbol=cross):
plot2 := plot(rhs(bestfit),x=0..16):
display({plot1, plot2});
```

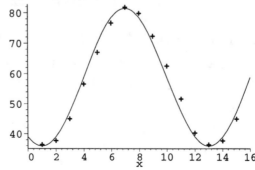

The temperature data fit nicely in this model. (The result can be even more impressive if we choose $a_4 = 7.25$ instead. Try it yourself!)

**A U.S.
Population
Model**

■ **Example.** The U.S. population (measured in millions) over the last two hundred years, between 1790 and 1990, is given in ten increments as follows:

```
year :=[seq(1790 + i*10, i=0..20)];
```

$year := [1790, 1800, 1810, 1820, 1830, 1840, 1850, 1860, 1870, 1880,$
  $1890, 1900, 1910, 1920, 1930, 1940, 1950, 1960, 1970, 1980, 1990]$

```
population := [ 3.929, 5.308, 7.240, 9.638, 12.861,
    17.063, 23.192, 31.443, 38.558, 50.189, 62.980,
    76.212, 92.228, 106.021, 123.203, 132.166, 151.326,
    179.323, 203.302, 226.542, 248.710]:
```

Theoretically, this population growth will follow a "logistic model" and hence has the form:

$$\text{population} = \frac{288.5}{1 + e^{a+bx}}, \text{ where } x \text{ is the year}$$

(The constant 288.5 represents the maximum sustainable population 288.5 million, which can be predicted by the data.) We want to find the constants a and b so that the logistic model best fits the data.

We cannot use the **leastsquare** command directly to find a and b. However, you can change the expression as follows:

$$\text{population} = \frac{288.5}{1 + e^{a+bx}} \quad \Leftrightarrow \quad \frac{288.5}{\text{population}} = 1 + e^{a+bx}$$

$$\Leftrightarrow \quad \ln(\frac{288.5}{\text{population}} - 1) = a + bx$$

This suggests that we can find a and b using the new data $\ln(\frac{288.5}{\text{population}} - 1)$. Thus, we must transform each of the given data pairs $[x, y]$ to be $[x, \ln(\frac{288.5}{y} - 1)]$.

```
poptrans := map(y->ln(288.5/y-1),population);
```

$poptrans := [4.2825978425, 3.9769099964, 3.6596583020, 3.3650034176,$
  $3.0648925623, 2.7668176917, 2.4370840236, 2.1011214673,$
  $1.8690652898, 1.5577806642, 1.2755916312, 1.0244249128,$
  $.75523768025, .54299791023, .29391045440, .16793603591,$
  $-.098186257082, -.49621823061, -.86971459877, -1.2964736219,$
  $-1.8326719347]$

(The **map** command will be explained in Chapter 24.)

```
leastsquare[[x,y]]([year,poptrans]);
```

$$y = 55.324474840 - .028552940449\, x$$

This suggests that $a = 55.3245$ and $b = -0.0285529$. The logistic model will thus be:

```
g := x -> 288.5/
        (1 + exp(55.32447484 - .028552940449*x)):
with(stats[statplots]):
pict1 := scatterplot( year, population):
pict2 := plot(g(x),x=1790..1990):
display({pict1, pict2});
```

(Use **scatter2d** instead of **scatterplot** in Release 3 or 4.)

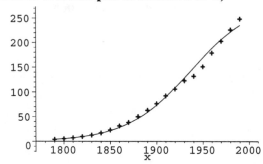

The logistic model fits the data very nicely, especially through about 1930.

Useful Tips

💡 💡 If the form $a_1f_1(x) + a_2f_2(x) + ... + a_nf_n(x)$ is omitted from the **leastsquare** command, *Maple* will fit a linear equation.

💡 💡 The **leastsquare** command will fit a linear equation with two or more variables. Check the examples given in the on-line help with **?leastsquare**.

💡 If you are planning to do a lot of statistics, we recommend that you also read Chapters 24 and 25, so you can learn more about lists, random numbers, and simulation.

Troubleshooting Q & A

Question ... I used the **leastsquare** command, but got error messages such as "... requires data of type stats," or "... unassigned matrix element," or "... cannot be used to initialize the range." What went wrong?

Answer ... Your input data has to be a list of statistical lists. In other words, it should be of the form,

[*list1*, *list2*, ..., *listn*]

where all the lists have the same number of entries. The first error message indicates your data is not a list of lists. The next two error messages indicate that the lists have different lengths.

Question ... When I tried to graph the points of a data set along with the polynomial obtained from using **interp**, the curve did not go through the points. What went wrong?

Answer ... The inaccuracy comes from the roundoff error when **interp** calculates the coefficients of the polynomial. You need to increase the number of digits used in the floating point computation, for example, try:

```
Digits := 12;
```

Question ... Can you explain the difference between using the **leastsquare** and **interp** commands?

Answer ... **leastsquare** finds a function of a specified form that best *approximates* the given set of data. On the other hand, **interp** will find a function that *matches the data exactly*, although it may behave "wildly." See the second example of the "More Examples" section.

CHAPTER 23

Animation

Getting Started

The animate Command

It is fun and easy to do animation in *Maple*. For 2-D animations, you use the **animate** command in the following form (it is defined in the **plots** library):

```
with(plots);
animate( a function of t and x, x = x_0 .. x_1, t = t_0 .. t_1 );
```

The *function* inside **animate** can be any object that will work in the **plot** command that we discussed in earlier chapters. For example, type

```
with(plots):
animate( sin(t*x), x = 0..2*Pi, t = 1..10);
```

Now, click on the picture, then click the play button ▶ on the control strip to see the animation.

In Release 4 or 5, the control strip will show up near the top of the worksheet after you click at the picture. It has the control buttons as that of a VCR.

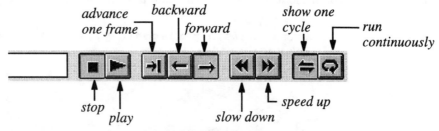

In Release 3, the control strip appears on the bottom of an animation window.

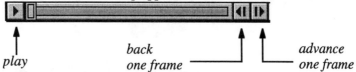

How does the Animation work?

In the example above, when you type:

```
animate( sin(t*x), x = 0..2*Pi, t = 1..10);
```

Maple first produces frames of the animation for numerous values of *t* between 1 and 10. These frames are approximately the same as the output of the commands:

```
plot(sin(1*x),x = 0..2*Pi);      # frame 1.
plot(sin(1.6*x),x = 0..2*Pi);    # frame 2.
# etc.
```

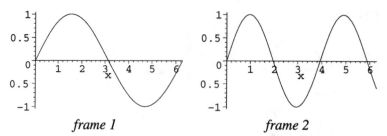

frame 1   frame 2

After *Maple* finishes creating all the pictures (sixteen by default), it then displays them rapidly enough so that your eye sees a continuous motion. This is how the animation is created.

The **animate** command works the same way as the **plot** command. The following demonstrate the various forms of syntax you can use with the **animate** command.

```
animate({t*x, -t*x, t*x^2, -t*x^2, t*x^3, -t*x^3},
        x=-1..1, t=-4..4);
```

(Here, each frame consists of the graphs of tx, $-tx$, tx^2, $-tx^2$, tx^3 and $-tx^3$.)

```
animate( {[t*cos(x), t*sin(x), x=0..2*Pi],
    [(1-t)*cos(x)+2, (1-t)*sin(x), x=0..2*Pi]}, t=0..1,
    scaling = constrained);
```

(In this case, each frame consists of the two parametric circles $(t\cos x, t\sin x)$ and $((1-t)\cos x+2,\ (1-t)\sin x)$.)

Animating a 3-D Picture

You can make 3-D animations in *Maple* as well. You simply use the **animate3d** command just as you used the **animate** command. For example,

```
with(plots);
animate3d(sin(t*x)+cos(t*y),x=0..Pi, y=0..Pi,t=1..5,
    axes=boxed);
```

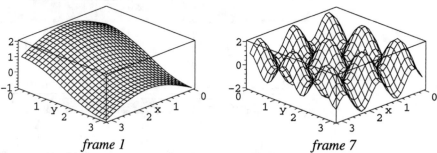

frame 1   frame 7

We show only two frames of the animation above (frames 1 and 7). If you watch the animation, you will see "waves" move through the surface.

Advanced Examples

The display Command

The examples we have looked at so far have been simple animations in which each frame could be plotted with a single plot command.

When we want to make more complicated animations, we must make double use of the **display** (or **display3d**) command, once to make the frame, then a second time to provide the animation.

An Animation Strategy

The key to writing sophisticated animations is to have a firm grasp of what each frame looks like, as well as how the frames vary according to a parameter. Following this concept, you can easily create a complicated animation.

In each of the following examples, we'll define **oneFrame**, a "frame function" of t, which will generate the graphic elements for each frame. Because each frame itself will be a combination of several simple graphics, we'll use the **display** command.

```
oneFrame := t ->display([ graphics command(s) depending on t ]);
```

Then we can use the **display** command with a sequence of frames. This time we set the **insequence** option to **true** to provide animation.

```
with(plots);
display([ a sequence of frames ], insequence=true);
```

Moving a Point Along a Curve

■ **Example**. Let's show you how to write the animation of a point moving along the sine graph, $y = \sin x$. The coordinate of the point at any time t can be thought as $P(t) = (t, \sin(t))$ for $0 \le t \le 2\pi$.

In each frame of the animation, we will plot the position of the point at time t against a background of the sine curve. The sine curve can be sketched with a **plot** command, and the point can be drawn using a **pointplot** command. We combine them using **display**.

```
oneFrame := t -> display([ plot(sin(x),x=0..2*Pi),
       pointplot([[t, sin(t)]], symbol=circle)]):
```

For example, you can see the frame at $t = 0.5$ with:

```
oneFrame(0.5)
```

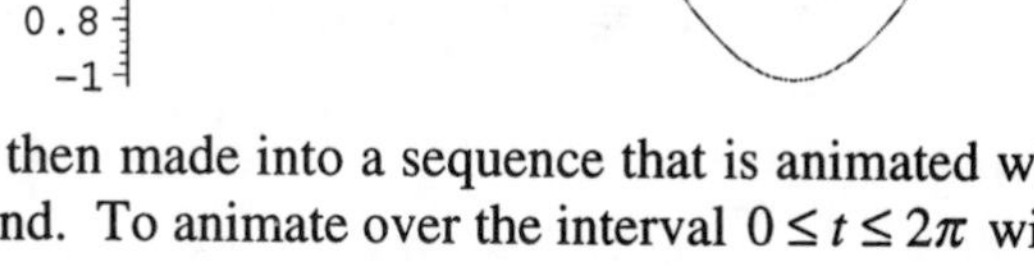

The frames are then made into a sequence that is animated with a second use of the **display** command. To animate over the interval $0 \le t \le 2\pi$ with 15 *frames* we use:

```
display([seq(oneFrame(i*2*Pi/14),i=0..14))],
    insequence=true);
```

Rolling a Ball Along the Ground

■ **Example**. A circle of radius one "rests" on top of the x-axis at the origin, as pictured below. A point P, which is marked at the "top" of the circle, is currently touching the y-axis at the point $(0, 2)$.

As the circle begins to "roll" to the right, the point P will rotate downward and eventually hit the x-axis after the circle has rolled a distance of π (because P was originally halfway around a circle of circumference 2π). Thereafter, P will rotate upward again after the circle "rolls over" it, reaching a high point again when the circle has rolled a distance of 2π.

The following picture suggests the motion of P.

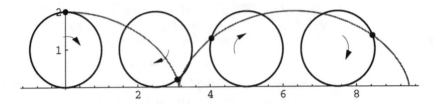

In fact, the path followed by P as the circle rolls is known to be a cycloid given parametrically by $(u + \sin(u), 1 + \cos(u))$, for $-\infty < u < \infty$.

We will let the parameter t in this animation be the distance the ball has rolled. Each frame of the animation should have three elements:

- A fixed portion of the cycloid to act as a background. It can be given by $(u + \sin(u), 1 + \cos(u))$, for $0 \leq u \leq 4\pi$.

- The new position of the unit circle in this frame. Since the circle has moved a distance t, its center is at $(t, 1)$ and its equation is thus $(t + \cos(u), 1 + \sin(u))$.

- The corresponding position of the point P. After the ball has traveled a distance of t, P will have rotated a distance of t clockwise around the circle. P's location relative to the center of the circle is thus given by $(\sin(t), \cos(t))$. Its position in the frame will then be $(t + \sin(t), 1 + \cos(t))$.

After some experimentation with **view**, our frame function is as follows.

```
oneFrame := t -> display([
    plot({[u+sin(u),1+cos(u),u=0..4*Pi],
          [t+sin(u),1+cos(u),u=0..2*Pi]}),
    pointplot([[t+sin(t), 1+cos(t)]], symbol=circle)],
    view=[-1..10,-.1..2.1], scaling=constrained):
```

The completed animation which rolls P over two arcs of the cycloid is given with:

```
display([seq(oneFrame(i*4*Pi/16),i=0..16)],
    insequence=true, scaling = constrained);
```

The Oscillating Spring

■ **Example**. A metal plate is attached to a spring of height h_0. (The picture to the right shows such a spring at a height of 5 units above the xy-plane.) The plate is then displaced upward a distance of c units and released. The plate (and spring) will begin a vertical, damped oscillation.

If we ignore the effect of gravity and the weight of the plate, then the height of the plate at any time t after release will be

$$z = h_0 + c\,e^{-bt}\cos(kt),$$

where b and k are constants that depend on the spring.

We'll model the motion of the plate using animation in *Maple*. The graphic elements of each frame are as follows.

- The height of the "plate" is $z = h_0 + c\,e^{-bt}\cos(kt)$. It can be seen with **plot3d**. We'll sketch it only over the rectangle $-1 \le x \le 1$, $-1 \le y \le 1$.

- We'll think of the "spring" as the helix $(\cos(u), \sin(u), \frac{u}{8\pi}(h_0 + c\,e^{-bt}\cos(kt))$, for $0 \le u \le 8\pi$, and sketch it using **spacecurve**.

Here's the frame function:

```
oneFrame := t -> display([
    spacecurve([cos(u), sin(u),
                u*(h0+c*exp(-b*t)*cos(k*t))/(8*Pi)],
        u=0..8*Pi, axes=boxed),
    plot3d(h0+c*exp(-b*t)*cos(k*t), x=-1..1, y=-1..1)]):
```

Before you can see an animation, you must choose numerical values for the various constants h_0, c, b, and k we've introduced. For example:

```
h0 := 5;      # The plate is at height 5 at rest.
c  := 2;      # It is displaced 2 units upward.
b  := 0.05;   # This coefficient controls damping.
k  := 1;      # This coefficient controls speed of oscillation.
```

Now you can do the animation, say, over the time interval $0 \le t \le 8$:

```
display([seq(oneFrame(i*8/16),i=0..16)],
        insequence=true);
```

Try repeating the animation for different numerical values of the constants and see how each one affects the movement of the plate. For example, increasing the value of b will cause the oscillation to die out more quickly.

Useful Tips

The **plot** command has many options. Some, like **style**, apply to the object being graphed. Others, like **titlefont**, apply to background settings.

However, the **display** command cannot reconcile conflicting background settings from the various plot structures that make up an animation. It simply chooses some and ignores others. Adding *identical* options like **axes**, **title**, and **tickmarks** to each plot structure will help you control the output of an animation.

Troubleshooting Q & A

Question...   Neither the **animate** nor **display** command seemed to work. What should I check?

Answer...   Make sure you have used the **with(plots);** command earlier in your session.

Question...   How do I decide whether to use **animate** or **display**?

Answer...   Since **animate** is simpler and easier to use than the double use of the **display** command, use **animate** and **animate3d** whenever you can.

You can use **animate** and **animate3d** whenever you can graph a single frame with the **plot** or **plot3d** command. This will be the case when you simply want to animate the graph of a function, or even several functions, with a time parameter.

If you need other graphing commands, like **pointplot**, **implicitplot**, or **spacecurve**, to graph a single frame, then you need to use the **display** command.

Question...   I tried using the **display** command, but I got pages of numbers instead of a picture. What happened?

Answer...   When the **display** command fails, it fails spectacularly. It shows the input it has received, which is a series of plot objects. Each plot object is a long sequence of numbers that tells *Maple* what it is trying to plot, including all the points on the graph and all the tickmarks on the axes.

The animations we have been looking at are compound plot structures. If the animation doesn't work, check the pieces.

Use a command like **display([oneFrame(1)]);** to test if individual frames graph properly. If the frames graph properly, the difficulty is with the syntax of the **display** command that does the animation. The most common errors are misspelling a keyword and forgetting to use [*square brackets*] to enclose the sequence of frames.

If the problem is with an individual frame, try plotting the elements individually for a variety of values of the parameter *t*.

Question...   *Maple* ran out of memory before completing the animation. What should I do?

Answer...   Animations use lots of memory (especially in 3-D).

Try increasing the memory allocation for *Maple* . Closing unneeded windows and quitting other applications before restarting may help too.

Also, try your animation using a small number of frames. Gradually increase the number of frames until you get a satisfactory result.

CHAPTER 24
More About Lists

In this chapter we will show you more commands that you can use when you work with lists. Many of them will be helpful in statistics or *Maple* programming.

Basic Lists

Setting Up and Reading Lists

We have been using lists throughout this text. (See chapter 6.) At a basic level, a list is an ordered sequence enclosed in square brackets. The entries of a list can be any valid *Maple* object.

For example, consider the list,

```
myList := [`red`, 2, 3, 3.153, x^2=5, sin(z),
           Pi, `history`, Fred];
```

$$myList := [\ red, 2, 3, 3.153, x^2 = 5, \sin(z), \pi, history, Fred\ ]$$

Elements of **myList** include strings, a name, integers, real numbers, an equation, and a function. Each element can be recalled by **myList[1]**, **myList[2]**, and so on.

```
myList[5];
```

$$x^2 = 5$$

The syntax **myList[–n]** denotes the element of **myList** that's *n* entries from the end of the list.

```
myList[-3];
```

$$\pi$$

We can use **myList[a..b]** to specify a range of entries of **myList**.

```
myList[3..6];                    # Shows the third to sixth elements.
```

$$[3, 3.153, x^2 = 5, \sin(z)]$$ # Result from Release 4 or higher.

$$3, 3.153, x^2 = 5, \sin(z)$$ # Result from Release 3 or lower.

Note that the result of **myList[a..b]** is different between versions of *Maple*. Release 4 or higher will give the result in a list structure.

If you want to see the elements without a list, then you use the **op** command:

```
op(myList[3..6]);                          # For Release 4 or higher.
```

$$3, 3.153, x^2 = 5, \sin(z)$$

(The command **op(myList[3..6])** in Release 4 or higher will have the same effect as the command **myList[3..6]** in Release 3.)

If an element of a list is a list itself, we can retrieve its entries in the same manner, using [*square bracket*] syntax. For example,

`newList := [[1, 5, 7], [-6, 5, 7,7], 4, 6, 8]:`

`newList[2];`   # The second element of **newList** is a list itself.

$$[-6, 5, 7, 7]$$

`newList[2][1..3];`   # We can also find the elements of an element.

$$[-6, 5, 7]$$ # Result from Release 4 or higher.

Modifying the Elements of a List

To change an entry in a list, you can use a normal assignment statement. Recall the list that we used earlier.

`myList;`

$$[red, 2, 3, 3.153,\ x^2 = 5,\ \sin(z),\ \pi,\ history,\ Fred]$$

We want to replace the first element of this list, say, with **green**:

`myList[1] := green;`   # For Release 4 or higher.

$$myList_1 := green$$

Now you can see that the first element of **myList** has been replaced.

`myList;`

$$[green, 2, 3, 3.153,\ x^2 = 5,\ \sin(z),\ \pi,\ history,\ Fred]$$

We can also modify a list directly by redefining its elements:

`myList := [op(myList[1..3]), good, bad, ugly,`
`                op(myList[5..7]), 100, 101];`

$$myList := [green, 2, 3,\ good,\ bad,\ ugly,\ x^2 = 5,\ \sin(z),\ \pi,\ 100, 101]$$

(Release 3 or earlier users will not need the **op** command above.)

The new **myList** consists of the first three elements of the original list, then the new elements "good, bad and ugly," followed by the fifth to seventh elements of the original list, and then the numbers 100 and 101.

As an example, it might be of interest to construct a list of the first 1000 prime integers, using the **ithprime** operator. We find these with:

`lotsOfPrimes := [ seq(ithprime(i), i=1..1000) ]:`

We can see the first three and last three elements of this list with:

`[op(lotsOfPrimes[1..3]), op(lotsOfPrimes[998..1000])];`

$$[2, 3, 5, 7901, 7907, 7919]$$

You were probably already familiar with the first few primes, but not many people realize that 7919 is the 1000th prime.

Useful List Commands

The map Command

To evaluate a function at each element of a list, you use the **map** command. It has the form:

> **map (** *function name* , *the list* **);**

Let's look at an easy example:

> **list1 := [seq(2^n, n=1..8)];**
>
> $list1 := [2, 4, 8, 16, 32, 64, 128, 256]$

> **f1 := x -> x+1 :**   # We want to add 1 to each element.
> **map(f1, list1);**
>
> $[3, 5, 9, 17, 33, 65, 129, 257]$

> **Note:** Inside the **map** command, we type the *name* of the function only. In the example above, we use **f1** and not **f1(x)**.

Now, we'll try a more interesting example:

> **list2 := [[`John`, 75, 62], [`David`, 62, 81],**
> **[`Mary`, 75, 91], [`Jane`, 31, 50],**
> **[`Steve`,21, 31]]:**

This shows a list of five students and their scores in two exams. We will calculate the average score of each student and list it with their name:

> **f2 := x -> [x[1], (x[2]+x[3])*0.5]:**

(Note that **x[1]** is the name of the student, while **x[2]** and **x[3]** are the exam scores.)

> **list2a := map(f2,list2);**
>
> $list2a := [[John, 68.5], [David, 71.5], [Mary, 83.0], [Jane, 40.5], [Steve, 26.0]]$

The **map** command provides an easy way to implement "data transformation" or "data massaging" if you work with statistical data. For an example, consult Chapter 22's example on modeling the U.S. population.

Sorting Lists

The **sort** command will arrange the elements of a list in an increasing order. It has the syntax:

> **sort (** *list* **);**

For example,

> **list3 := [81, 70, 97, 63, 76, 38, 85, 68, 21]:**
> **sort(list3);**
>
> $[21, 38, 63, 68, 70, 76, 81, 85, 97]$

You can also specify any "Boolean function" on two variables as the sorting criterion. It has the syntax:

> **sort (** *list,* *Boolean function* **);**

For example, you can sort the **list3** above in descending order.

```
f3 := (a,b) ->           # This Boolean function will return true if the
       evalb(a > b):     # first element is larger than the second element.
sort(list3, f3);
```
$$[97, 85, 81, 76, 70, 68, 63, 38, 21]$$

This flexibility lets us sort lists whose elements are lists themselves. Consider the following example that sorts our list of students according to their second exam scores:

```
list2 := [ [`John`, 75, 62], [`David`, 62, 81],
           [`Mary`, 75, 91], [`Jane`,  31, 50],
           [`Steve`,21, 31]]:

f4 := (x,y) -> evalb(x[3] < y[3]):
```

(Note that in this function, **x** and **y** refer to a list [*name, exam1, exam2*]; thus **x[3]** and **y[3]** are the third elements, *exam2*.)

```
sort(list2, f4);
```
$$[[\textit{Steve}, 21, 31], [\textit{Jane}, 31, 50], [\textit{John}, 75, 62], [\textit{David}, 62, 81], [\textit{Mary}, 75, 91]]$$

Counting the Elements

The command **stats[transform, tally[count]]** shows how many times each item appears in a list. It has the syntax:

```
stats[transform, tally[count]]( your list );
```

For example,

```
list1 := [ 3, 3, 5, 6, 5, 9, 10, 5, 5, 1, 5, 1, 5];
```
$$\textit{list1} := [3, 3, 5, 6, 5, 9, 10, 5, 5, 1, 5, 1, 5]$$

```
stats[transform, tally[count]](list1);
```
$$[6, 9, 10, \text{Weight}(1, 2), \text{Weight}(3, 2), \text{Weight}(5, 6)]$$

This means that 6, 9, and 10 appear once, 1 appears twice, 3 appears twice, and 5 appears six times.

Other List Commands

Many other commands for lists will be useful, especially for programming in *Maple*. We will list some of them here.

Maple *Command*	*Explanation*

`max(op([ 2, 5, -3, 1.2 , 6.01, 7.5]));`
 7.5
 `min(op([ 2, 5, -3, 1.2 , 6.01, 7.5]));`
 -3
	max and **min** are used to find the largest and smallest elements of a list or set, respectively.

`nops([ 2, 5, -3, 1.2 , 6.01, 7.5]);` 6	**nops** tells you the number of elements in a list or set.

More Examples

Baseball Standings

■ **Example**. The following list represents the final standings of the American League Eastern Division Baseball teams for the 1997 regular season in the form [*team, wins, losses*]:

```
baseball := [ [`Bal`, 98, 64], [`Bos`, 78, 84],
   [`Det`, 79, 83], [`NY`, 96, 66], [`Tor`, 76, 86]]:
```

When you look at the sports page of a newspaper, you'll see that the team standings include the team's winning percentage. Also, teams with a higher winning percentage will be listed first. Let's show how to do this in *Maple*:

- We will first add the winning percentage to each element (team):

```
f1 := x -> [op(x[1..3]),      # Note that in this function the element x
     evalf(x[2]/162,3)]:      # is the sublist [team, wins, losses] .

list1 := map(f1, baseball);
```

   list1 := [[*Bal*, 98, 64, .605], [*Bos*, 78, 84, .481], [*Det*, 79, 83, .488], [*NY*, 96, 66, .593], [*Tor*, 76, 86, .469]]

- Next we will **sort** the list in descending order of the winning percentage.

```
f2 := (x,y) -> evalb(x[4] > y[4]):
list2 := sort(list1, f2);
```

   list2 := [[*Bal*, 98, 64, .605], [*NY*, 96, 66, .593], [*Det*, 79, 83, .488], [*Bos*, 78, 84, .481], [*Tor*, 76, 86, .469]]

- Finally, we will print the contents of each team line by line, just like you see it in a newspaper:

```
for i from 1 to 5 do op(list2[i][1..4]) od;
```

Bal,	98,	64,	.605
NY,	96,	66,	.593
Det,	79,	83,	.488
Bos,	78,	84,	.481
Tor,	76,	86,	.469

(We will explain the **for** and **do** commands in detail in Chapter 26.)

A Game of Bridge, Anyone?

■ **Example**. What happens if you want to play bridge but you don't have a deck of cards? We can ask *Maple* to simulate shuffling a deck of cards.

A deck consists of 52 cards, arranged into four suits named "Clubs," "Diamonds," "Hearts," and "Spades" of 13 cards each. Asking *Maple* to shuffle cards is mathematically the same as arranging the numbers from 1 to 52 randomly. We can use the **randperm** command defined in the **combinat** library to do this:

```
shuffle := combinat[randperm](52);
```

   shuffle := [17, 38, 46, 4, 48, 21, 26, 2, 8, 32, 22, 15, 30, 6, 40, 45, 49, 24, 27, 44, 29, 39, 43, 41, 31, 3, 42, 23, 35, 1, 14, 16, 37, 12, 47, 19, 28, 50, 36, 20, 10, 25, 11, 33, 51, 9, 5, 13, 18, 52, 7, 34]

Next we need to deal the cards into the four hands and sort them in each hand.

```
west  := sort([seq(shuffle[4*i+1],i=0..12)]);
north := sort([seq(shuffle[4*i+2],i=0..12)]);
east  := sort([seq(shuffle[4*i+3],i=0..12)]);
south := sort([seq(shuffle[4*i+4],i=0..12)]);
```

$$
\begin{array}{lll}
\textit{west} & := [8, & 10, & 17, & 18, & 28, & 29, & 30, & 31, & 35, & 37, & 48, & 49, & 51] \\
\textit{north} & := [1, & 3, & 6, & 9, & 12, & 21, & 24, & 25, & 32, & 38, & 39, & 50, & 52] \\
\textit{east} & := [5, & 7, & 11, & 14, & 22, & 26, & 27, & 36, & 40, & 42, & 43, & 46, & 47] \\
\textit{south} & := [2, & 4, & 13, & 15, & 16, & 19, & 20, & 23, & 33, & 34, & 41, & 44, & 45]
\end{array}
$$

Now we need to interpret the cards numbered 1 to 13 as "Spades," those numbered 14 to 26 as "Hearts," 27 to 39 as "Diamonds," and 40 to 52 as "Clubs."

```
suitnames := x ->   if   x <14   then   [`S`,x]
                    elif x<27    then   [`H`,x-13]
                    elif x<40    then   [`D`,x-26]
                    else  [`C`,x-39] fi:
```

```
west1  := map(suitnames, west):
north1 := map(suitnames, north):
east1  := map(suitnames, east):
south1 := map(suitnames, south);     # We will only show you
                                     # South's hand here.
```

$\textit{south1} := [[S, 2], [S, 4], [S, 13], [H, 2], [H, 3], [H, 6], [H, 7], [H, 10],$
$\qquad\qquad [D, 7], [D, 8], [C, 2], [C, 5], [C, 6]]$

It will help to interpret 1 as "Ace," 11 as "Jack," 12 as "Queen," and 13 as "King":

```
cardnames := x ->   if   x[2]= 1   then   [x[1],`Ace`]
                    elif x[2]=11    then   [x[1],`Jack`]
                    elif x[2]=12    then   [x[1],`Queen`]
                    elif x[2]=13    then   [x[1],`King`]
                    else [x[1], x[2]] fi:
```

Finally, we convert the sorted hands to the more familiar bridge form.

```
west2  := map(cardnames, west1);
north2 := map(cardnames, north1);
east2  := map(cardnames, east1);
south2 := map(cardnames, south1);
```

$\textit{west2} \quad := [[S, 8], [S, 10], [H, 4], [H, 5],$
$\qquad\qquad [D, 2], [D, 3], [D, 4], [D, 5], [D, 9], [D, \textit{Jack}], [C, 9], [C, 10], [C, \textit{Queen}]]$

$\textit{north2} \quad := [[S, \textit{Ace}], [S, 3], [S, 6], [S, 9], [S, \textit{Queen}], [H, 8], [H, \textit{Jack}], [H, \textit{Queen}],$
$\qquad\qquad [D, 6], [D, \textit{Queen}], [D, \textit{King}], [C, \textit{Jack}], [C, \textit{King}]]$

$\textit{east2} \quad := [[S, 5], [S, 7], [S, \textit{Jack}], [H, \textit{Ace}], [H, 9], [H, \textit{King}],$
$\qquad\qquad [D, \textit{Ace}], [D, 10], [C, \textit{Ace}], [C, 3], [C, 4], [C, 7], [C, 8]]$

$\textit{south2} \quad := [[S, 2], [S, 4], [S, \textit{King}], [H, 2], [H, 3], [H, 6], [H, 7], [H, 10],$
$\qquad\qquad [D, 7], [D, 8], [C, 2], [C, 5], [C, 6]]$

It looks like both North and East have pretty good hands.

Useful Tips

The internal representation of *Maple* lists is different for lists with more than 100 elements. With such long lists we cannot use subscript notation to change values of an element of a list. (For example, you cannot use **longlist[3]** := *new element*.) You may try **newlonglist** := [**op(longlist[1..2])**, *new element*, **op(longlist[4..200])**].

If order and repetition are not important, we are probably more interested in a *set* rather than in a list. You can then use set commands such as **union, intersection**, and **minus**. Recall that sets are written as sequences contained in { *curly braces* }.

CHAPTER 25

Random Numbers and Simulation

Random Numbers

The rand Command

We can use *Maple* to generate sequences of random numbers. Such sequences can be used to simulate probabilistic situations.

The following command will generate a random integer between given numbers *a* and *b*.

```
rand[ a .. b ]();
```

If you need a random integer between 0 and *n*–1, you can also enter:

```
rand[ n ]();
```

For example, you want a random integer between 1 and 6:

```
rand(1..6)();
            6
```

For a random integer between 0 and 9,

```
rand(10)();
            3
```

Pseudo-Randomness

Sequences of numbers returned by **rand** are said to be "pseudo-random." They are actually generated using an iteration formula. They eventually repeat but not soon enough so that you would notice.

And, as you see below, the values of **rand** certainly "look" random!

```
xdata := [seq(i, i=1..1000)] :
ydata := [seq(rand(100)(), i=1..1000)]:

with(stats[statplots]):  scatterplot( xdata, ydata);
```

(A thousand random numbers between 0 and 99 are generated and plotted as the *y*-coordinates.)

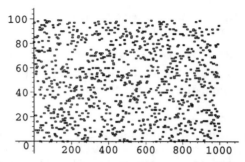

The fact that there seems to be no obvious pattern emerging confirms our sense of randomness.

■ **Example**. Let us generate 5000 random integers between 1 and 10:

```
data := [seq(rand(1..10)(), i=1..5000)] :
```

You can see some of them with:

```
[op(data[1..5]), op(data[4995..5000])];
```
$$[3, 3, 8, 1, 5, 7, 4, 7, 4, 2, 9]$$

If these values were truly random, we would expect each of the numbers 1, 2, . . . , 10 to show up about the same number of times. Let's check it:

```
stats[transform,tally[count]](data);
```

[Weight(1, 497), Weight(2, 490), Weight(3, 513), Weight(4, 500),
Weight(5, 489), Weight(6, 481), Weight(7, 520), Weight(8, 510),
Weight(9, 499), Weight(10, 501)]

So, 1 appeared 497 times, 2 appeared 490 times, and so forth. In fact, each digit appeared almost equally often (close to 500 times). Consequently, we say that **rand** produces a uniform distribution of numbers between 1 and 10.

Examples in Simulation

With the help of the **rand** command and a little programming, we can simulate many real-world experiments in *Maple*. In each of the following simulations, usually three procedures are involved.

- We will write down the procedure(s) that generates the result of a single experiment. The procedure is written in the form:

```
game := proc()
        local local variables separated by commas ;
        Maple commands separated by semicolons or colons ;
        end;
```

In many cases the procedure involves the **if** command, used in the form:

```
if condition then result1 else result2 fi;
```

Maple will give *result1* if *condition* is *true*. If *condition* is *false*, *result2* is returned. (The **proc** and **if** commands will be discussed in detail in the next chapter.)

- Then, we use the **seq** command to repeat the experiment a large number of times in the form:

  data := [seq(*one result of a random experiment,* **i = 1..*n*)];**

- After we obtain experimental data in this way, we'll analyze the data and summarize what we observed.

Let's Flip a Coin

■ **Example**. The flipping of a "fair" coin will give us either a head or a tail, with both results being equally probable. That is, approximately 50% of the time we'll get a head, and 50% of the time we'll get a tail. Since **rand(0..1)()** gives a random integer 0 or 1, we can simulate flipping a coin with:

```
flip := proc()  local a;
  a:= rand(0..1)();
  if a = 0 then `Head` else `Tail` fi;
end:
```

This guarantees that every time *Maple* evaluates **flip()**, you get a new random number and a new flip of the coin.

```
flip();
```
> *Head*

```
flip();
```
> *Tail*

Here are eight coin flips:

```
seq(flip(), i=1..8);
```
> *Tail, Tail, Tail, Head, Head, Tail, Head, Tail*

Let's Roll a Die

■ **Example**. A standard die has six sides labeled 1, 2, 3, 4, 5, and 6. If the die is "fair," we expect each number to appear on the top face about one-sixth of the time. We can simulate rolling a die with:

```
rand(1..6)();
```
> 5

```
rand(1..6)();
```
> 1

Now, suppose that players *A* and *B* each roll a die. The one who gets the larger number wins the game. We can simulate this game:

```
game := proc()
        local player1, player2;
        player1 := rand(1..6)():
        player2 := rand(1..6)():
            if player1 > player2 then `A wins`
            elif player1 = player2 then `Tie`
            else `B wins` fi;

        end;
```

Each time you evaluate the command **game()**, you get a new result.

```
game();
```

> *B wins*

```
game();
```

> *Tie*

Let us repeat this game 1000 times and count how many times each player wins.

```
data := [seq(game(), i=1..1000)]:
stats[transform,tally[count]](data);
```

> [Weight(*B wins*, 440), Weight(*Tie*, 167), Weight(*A wins*, 393)]

In this case, it seems that player *B* had better luck.

A Strange Behavior of Numerical Answers

■ **Example**. Look at the answer sections in your physics, chemistry, or economics textbooks. Do you notice that the first digit of each numerical answer is more likely to be 1, 2, 3, or 4 rather than 5, 6, 7, 8, or 9? This is because after each multiplication or division of two numbers, the first digit of the result is more likely to be 1, 2, 3, or 4 than the others. (Most scientific values arise from multiplications and divisions.)

This phenomenon is related to properties of the logarithm function. If you don't believe it, let's do an experiment to demonstrate it.

We will choose two integers randomly between 1 and 100, multiply them together, and record the first digit of the product. It is tricky to extract the first digit of the product. We will first **convert** the product into string, then use the **substring** command to isolate its first character, and use **parse** to extract the digit.

```
game := proc()
    local integer1, integer2, theProduct:
    integer1 := rand(1..100)():
    integer2 := rand(1..100)():
    theProduct := integer1*integer2;
    parse(
        substring(convert(theProduct, string),1..1));
    end;
```

Here are some test runs:

```
game();
```

> 1

```
game();
```

> 4

Now we repeat this experiment 5000 times.

```
data := [seq(game(), i=1..5000)]:
data1 := stats[transform,tally[count]](data);
```

data1 := [Weight(9,201), Weight(1,1149), Weight(2,955), Weight(3,712), Weight(4,587), Weight(5,470), Weight(6,399), Weight(7,302), Weight(8,225)]

The result of "which lead digit occurs most often" is easy to summarize, if we sort the data with **statsort**:

```
stats[transform,statsort](data1);
```

[Weight(1,1149), Weight(2,955), Weight(3,712), Weight(4,587),
Weight(5,470), Weight(6,399), Weight(7,302), Weight(8,225), Weight(9,201)]

Are you convinced?

A Game of Risk™, Anyone?

■ **Example.** In the Game of Risk™, players attempt to control a map of the world by occupying countries with their armies. In each turn of the game, a player (the "attacker") may choose to engage another player (the "defender") who occupies an adjacent country in battle. If the attacker eliminates the armies of the defender, the attacker takes over the country.

To simulate a battle, the attacker is allowed to roll two dice, but the defender only one. If the attacker's larger die is higher than the defender's die, the attacker wins the battle. However, the defender wins if his die is higher than or ties the larger die of the attacker.

The attacker has the advantage of rolling more dice, but the defender wins ties. So, which player has the advantage?

To find out, we can simulate the action of the attacker by rolling two dice and selecting the largest.

```
attacker := x -> max( rand(1..6)(), rand(1..6)() ):
```

The action of the defender is easier to simulate:

```
defender := x -> rand(1..6)():
```

A sample battle might look like this:

```
attacker();
```
                5

```
defender();
```
                4

In this case, the attacker wins.

The only outcomes of a battle can be that the attacker wins (+1) or loses (–1). Thus we can simulate a battle with:

```
battle := x -> if attacker() > defender() then 1
                   else -1 fi;
```

Now, we can conduct 1000 battles and summarize with:

```
data := [seq( battle(), i=1..1000)]:
stats[transform,tally[count]](data);
```
          [Weight(-1, 425), Weight(1, 575)]

The result clearly shows that the attacker has the advantage over time. (In fact, the theoretical probability of the attacker winning over the defender is $\left(\frac{5}{6}\right)^3 \approx 58\%$. Our simulation is quite representative at 575/1000 = 57.5%.)

Happy Birthday to YOU!!

■ **Example.** The "birthday problem" is a famous problem in probability. For example, in a room of 30 people, how likely is it that at least two of them have the same birthday? We'll suggest an answer using *Maple*.

It's sensible to assume that birthdays of 30 randomly assembled people are spread evenly over the year, so we can simulate by picking 30 birthdays at random from 365 days of the year (sorry, we don't do leap years in this experiment!)

```
room := proc() local i;
        seq(rand(1..365)(), i=1..30); end;

room();
```

    66, 97, 185, 58, 148, 212, 105, 280, 164, 289, 301, 83, 88, 20, 148, 109, 48, 166, 327, 214, 208, 168, 275, 3, 277, 233, 288, 308, 65, 1

The output above simulates a typical "room" of 30 people, represented by their birthdays. Notice for this output that the number 148 appears twice, indicating that two people of this group had the same birthday on the 148th day (May 28).

We can count how many *distinct* birthdays there are in the room with:

```
nops({ room() });
```

    29

(The command { **room()** } will include the birthdays in a set and hence *removes duplicates.* **nops**({**room()** }) is then the number of distinct birthdays.)

When **nops**({**room()**}) **< 30** is *true*, there's a duplicate birthday in the room. Now we can experiment with 1000 rooms and count the number of *true* results we see:

```
data := [seq(evalb( nops({room()}) < 30), i=1..1000)]:
stats[transform,tally[count]](data);
```

    [Weight(*false*, 298), Weight(*true*, 702)]

That is, the probability that there are two people with the same birthday in a room of 30 randomly assembled people is about 702/1000 = 70.2% Are you surprised? Nearly everyone is surprised the first time they see this result. In fact, the theoretical probability that at least two people have the same birthday in a room of 30 people is approximately 70.6%. Our simulation was very close.

CHAPTER 26

Maple for Programmers

This chapter is designed for *Maple* users who are already familiar with writing computer programs. However, it is not a tutorial on programming. Rather, we expect that you're looking for information on those areas of *Maple* which are closely related to programming concepts. This chapter is divided into the following sections:

- Traditional programming languages
- Recursive structures in *Maple*
- Code generation for FORTRAN or C
- Viewing *Maple* code
- Importing and Exporting data

Elements of Traditional Programming Languages

Maple has its own programming language that supports elements found in traditional computer programs: conditional execution, looping, and subroutines. The syntax of the language borrows from the Pascal and C programming languages.

> **Note:** Until now, we have used the term "command" to describe *Maple* syntax (e.g., we used the **plot3d** command in Chapter 15). In this chapter, most commands will be called **statements**. This is more consistent with the terminology of programming languages.

The print Statement

To print the value of one or more expressions, use the statement:

```
print( expression1 , expression2 , ... );
```

The **print** statement is useful when you are developing programs and are trying to debug them.

The if Statement

The most commonly used form of the **if** statement is:

```
if logical condition then result1; else result2; fi;
```

(Notice that the **if** statement ends with **fi**, a backwards **if**.) If the *logical condition* evaluates as *true*, *result1* is returned. If the *logical condition* evaluates as *false*, *result2* is returned. For example:

```
a := 4;  if  a <= 3 then 2*a-1; else 5*a+1; fi;
                21

a := 3;  if  a <= 3 then 2*a-1; else 5*a+1; fi;
                5
```

In an **if** statement, both *result1* and *result2* can be either a single statement or a collection of statements separated by semicolon(s). The logical decision can also have more than two branches by using the optional **elif** clause (for "else if"):

```
if logical condition then result1;
   elif logical condition then result2;
   else result3; fi;
```

Consider, for example, a procedure to solve the quadratic equation $ax^2 + bx + c = 0$:

```
a := 4:    b := 7:    c := 2:
disc := b^2 -4*a*c:
if disc > 0 then
   print(`The equation has two solutions at`);
   print((-b+sqrt(disc))/(2*a),` and `,
        (-b-sqrt(disc))/(2*a));
elif disc = 0 then
   print(`The equation has a double solution at`);
   print((-b)/(2*a));
else
   print(`The equation has no real solutions`);
fi:
```

The equation has two solutions at

$$-\frac{7}{8} + \frac{1}{8}\sqrt{17}, and, -\frac{7}{8} - \frac{1}{8}\sqrt{17}$$

Do Loop

The simplest kind of loop is the **do** statement. The syntax is:

```
do body od:
```

(**od** is **do** backwards.) The *body* can be either a single *Maple* statement or a sequence of statements separated by semicolons. To keep the loop from executing forever. you can use the **break** statement within the *body* of the **do** statement.

■ **Example.** The **nextprime** command can be used in a **do** loop to make a short list of primes:

```
aprime := 2;
do
   aprime := nextprime(aprime);
   if aprime > 12 then break fi;
od;
```

```
aprime := 2
aprime := 3
aprime := 5
aprime := 7
aprime := 11
aprime := 13
```

Using a **do** loop with one or more **break** commands to stop repetition is rather inelegant. The **for** loop provides more control over repetition of the loop.

For Loop

The most common **for** loop has the form:

```
for index variable from starting value to stopping value
do body od:
```

■ **Example.** The Fibonacci numbers are $c_0 = 1$, $c_1 = 1$, and $c_n = c_{n-1} + c_{n-2}$, for $n \geq 2$. Executing the assignment **c[n] := c[n–1] + c[n–2]** repeatedly for $n \geq 2$ computes these directly:

```
c:= array(0..25);
c[0] := 1;
c[1] := 1;
for i from 2 to 25 do
   c[i] := c[i-1] + c[i-2];
   od:
```

To see a few of the Fibonacci numbers, you can use:

```
print(seq(c[i],i=0..10));
```
$$1, 1, 2, 3, 5, 8, 13, 21, 34, 55, 89$$

```
c[21];
```
$$17711$$

Extensions of For Syntax

The general form of the **for** statement is:

```
for index variable (from expression)
    (by expression) (to expression)
    (while expression ) do body od;
```

where anything in parentheses is optional. If **from** and **by** are omitted, the default values of **from** 1 and **by** 1 are used. The **to** expression and **while** expression are tested at the beginning of each loop. Both can be used at the same time. The index variable is increased using the **by** value at the end of the loop.

■ **Example.** To compute the sum of the positive integers 1, 2, 3, . . ., 50, use:

```
intsum := 0:
for i from 1 to 50 do intsum := intsum + i od:
intsum;
```
$$1275$$

To compute the same sum using the **while** statement, use:

```
intsum := 0:
for i from 1 while i <= 50 do intsum := intsum + i od:
intsum;
```
$$1275$$

Procedures, Functions, and Subroutines

Almost every programming language supports a notion of a subroutine, function, or procedure. For example, BASIC allows the use of **DEF**, **GOSUB**, and **RETURN** statements, while FORTRAN has a **CALL** statement.

Subroutines are best written in *Maple* using the **proc** statement. Its general form is:

```
proc( parameter list)
    local    local variable list;
    global   global variable list;
    options  options list;
    body
    end;
```

Maple will execute the statement(s) in the *body* and then return the value of the last statement executed. Here are two examples.

■ **Example.** To compute the area of a triangle with sides having length a, b, and c, you can use Heron's Formula which says that $area = \sqrt{s(s-a)(s-b)(s-c)}$, where $s = (a+b+c)/2$ is the semi-perimeter of the triangle. A nice coding of an area function (subroutine) would be:

```
area := proc(a,b,c)
        local s;
        s := (a+b+c)/2;
        sqrt(s*(s-a)*(s-b)*(s-c));
        end:

area(5, 5 ,8);
        12
```

> **Note:** Don't forget to add semicolons or colons between statements in the body of a **proc**. Otherwise, *Maple* stops with a syntax error. Remember that colons suppress output of a statement.

Since the length of each side should be positive, we want the procedure to test if the input parameters are positive.

```
area := proc(a::positive,b::positive,c::positive)
        local s;
        s := (a+b+c)/2;
        sqrt(s*(s-a)*(s-b)*(s-c));
        end:
```

■ **Example.** We can write a routine that computes the maximum, minimum, and average of a list of exam scores from a Calculus class, and then reports the scores in increasing order.

```
summarize := proc(data)
  local sortedList, n:
  sortedList := sort(data):
  n := nops(data);
  print(`The number of students is `, n);
  print(`The maximum score is `, sortedList[n]);
  print(`The minimum score is `, sortedList[1]);
  print(`The average is `,
        evalf(sum(data[i], i=1..n)/ n, 4));
  sortedList;
end:

class1 := [61,23,14,78,91,33,12,44,72,79,82,81];
summarize(class1);
```

> *The number of students is* 12
> *The maximum score is* 91
> *The minimum score is* 12
> *The average is* 55.83
> [12, 14, 23, 33, 44, 61, 72, 78, 79, 81, 82, 91]

Recursive Structures in *Maple*

Maple allows a procedure to call itself. Thus recursive computations are done quite naturally.

■ **Example.** The Fibonacci numbers are defined recursively as $c_0 = 1$, $c_1 = 1$, and $c_n = c_{n-1} + c_{n-2}$, for $n \geq 2$. These can be defined in *Maple* with:

```
c := proc(n::nonnegint)
    if n < 2 then 1
    else c(n-1) + c(n-2) fi end:
```

The procedure call checks that **n** is a nonnegative integer. The condition "**if n < 2 then 1**" will define **c(0) := 1** and **c(1) := 1**. To find the Fibonacci number c_5, just type:

```
c(5);
```
$$8$$

To evaluate **c(5)**, *Maple* repeatedly uses the three rules above to simplify the expression in approximately the following sequence of transitions:

$$c_5 \rightarrow c_4 + c_3 \rightarrow (c_3 + c_2) + (c_2 + c_1) \rightarrow ((c_2 + c_1) + (c_1 + c_0)) + ((c_1 + c_0) + 1)$$
$$\rightarrow (((c_1 + c_0) + 1) + (1 + 1)) + ((1 + 1) + 1) \rightarrow (((1 + 1) + 1) + 2) + (3) \rightarrow 8$$

Remembering Values During Recursion

One disadvantage of the method above is that *Maple* will not remember the values that it has calculated recursively. For example, if you want to calculate **c(6)**, *Maple* recomputes **c(5)**, **c(4)**, **c(3)**, and **c(2)** all over again. This can slow down computation dramatically even for just, say, **c(20)**.

The **remember** option of *Maple* gives you a way to evaluate **c(5)** and at the same time define its value in case you need to use it later. You do this with the following new definition of the Fibonacci procedure:

```
newFib := proc(n::nonnegint)
    option remember:
    if n < 2 then 1
    else newFib(n-1) + newFib(n-2) fi end:
```

This option makes *Maple* create a "remember table" so that **newFib(i)** is only computed once. On subsequent calls **newFib(i)** is evaluated with a lookup table.

The effect on execution time is dramatic. Evaluating **c(20)** using the first recursive method above takes *two to three hundred* times longer than **newFib(20)** of the second method!

Code Generation

After we have worked with a procedure in *Maple*, we may want to use the results in a different computer program. The *Maple* commands **fortran** and **C**, respectively, pro-duce FORTRAN or C code equivalents of expressions and procedures. Both of these commands need to first be read in with **readlib**.

■ **Example.**   The procedure **newFib** above for finding Fibonacci numbers includes an error checking option that is not available in either FORTRAN or C. We modify the procedure slightly and then let *Maple* produce code for us.

```
newFib := proc(n)
   option remember:
   if type(n,negint) then
        ERROR(`bad arguments type`)
   elif n < 2 then 1
   else newFib(n-1) + newFib(n-2) fi end:

readlib(C):
C(newFib);
```

```
/* The options were    : remember */
double newFib(n)
double n;
{
    if (type(n,negint))
        {
            fprintf(stderr, "bad arguments type" );
            exit(1);
        }
    else
        if (n < 2)
            return(1);
        else
            return(newFib(n-1)+newFib(n-2));
}
```

```
readlib(fortran):
fortran(newFib);
```

(We will not show you the output here.)

Both of these commands have options to optimize the code and save the results directly to a file.

Viewing *Maple* Code

Viewing *Maple* Procedures

Almost all of the code in *Maple* can be viewed on demand. You may want to do this to see exactly how *Maple* executes a command, or because you want to modify or extend a procedure.

To view the *Maple* code in most procedures you need to change the value of the system variable **verboseproc** from 1 to 2 with the **interface** command.

```
interface(verboseproc=2);
```

Now, when you ask *Maple* to print a command it will give you the code the command calls up. For example to see the code for the command **norm**:

```
print(norm);
```

```
proc(p, n, v)
local c;
option `Copyright (c) 1992 by the University of
        Waterloo. All rights reserved.`;
    c := traperror(coeffs(p, args[3 .. nargs]));
    if c = lasterror then ERROR
                    (`polynomial must be expanded`)
    elif n = infinity then max(op(map(abs, [c])))
    elif n = 1 then convert(map(abs, [c]), `+`)
    elif n = 2 then,
        `norm/ex`(convert(map(x -> abs(x)^2,
                    [c]), `+`), 1/2)
    elif type(n, 'numeric') and 1 <= n then `norm/ex`(
        convert(map((x, n) ->
                    abs(x)^n, [c], n), `+`), 1/n)
    else ERROR(`norm not implemented`)
    fi
end
```

You should be warned that most common commands are composed of several pages of code, as *Maple* carefully breaks most commands into many cases.

File I/O

Saving Values for Future *Maple* Use

You can **save** results from one session of *Maple* to be **read** in during a future *Maple* session. Both commands use a pathname that starts in the same directory as the *Maple* application. The statement

```
save(`mondayMaple.m`);
```

saves the current value of all variables and procedures in a file named **mondayMaple.m**. The file would be read back in during a later *Maple* session with

```
read(`mondayMaple.m`);
```

This is an internally formatted *Maple* file and is not readable with a text editor. It is, however, in ASCII so that it can be transferred via e-mail.

Importing and Exporting with Other Programs

The quickest way to write to a file is with the **writeline** command. The command

```
writeline(myfile, string1 , string2 , ... );
```

will append all the string variables listed to the file *myfile*. Each string will be written on its own line. For example,

```
writeline(myFile1, `This is a testing`, `1`,`2`,`3`);
```

Now, if you open **myFile1** using a text editor, you will see the following lines.

```
This is a testing
1
2
3
```

To write numeric data, you must go through a three-step process of opening a file, writing to it, and closing the file.

- First, you must create file descriptor for the output file. For a buffered file, you use the **fopen** command in the form:

fopen(*file name*, *READ/WRITE/APPEND*, *TEXT/BINARY* **);**

For example, if we want to append text to **file1**, we would use the statement:

```
fd1 := fopen(file1, APPEND, TEXT);
```

- Next, you use the **writedata** statement to output the values. You may write a list or vector or matrix. To write **myData** to **file1**, use the statement:

```
myData := [20, 31, 15, 26, 17, 18];
writedata(fd1, myData);
```

- Finally, when you are finished with output, you should **fclose** the file:

```
fclose(fd1);
```

If you use a text editor to open **file1**, you will see the data 20, 31, etc., one at a line. You can read this file (or any data file) back to *Maple* using the **readdata** command.

```
readdata(file1);
```
$$[20, 31, 15, 26, 17, 18]$$

Optionally, you can add formatting information after the data. More than one **writedata** statement can be used between the **fopen** and **fclose**.

■ **Example.** Suppose that you want to compute the values of the Fibonacci numbers with the procedure **newFib** defined above, and then save those values to a file named "Fibonacci.dat", first with all 25 values on one line, then with 25 lines each containing the list [**i, newFib(i)**].

```
fd1 := fopen(Fibonacci.dat,WRITE,TEXT):
writedata(fd1,[[seq(newFib(i),i=1..25)]]);
for i from 1 to 25 do
    writedata(fd1, [[i,newFib(i)]]);
od;
fclose(fd1);
```

To read the data we can use either **readline** or **readdata**. The statement

```
readdata(Fibonacci.dat,3);
```

will try to read three numeric values from each line of **Fibonacci.dat**. The statement

```
readline(Fibonacci.dat);
```

will read the next line of **Fibonacci.dat**.

APPENDIX A
Worksheets (for Release 4 or Higher)

Worksheets and Groups

Worksheets
On every graphically based computer system, when you start *Maple* you will get a window that consists of a **menu bar**, a **tool bar**, a **context bar**, a **worksheet**, and a **status bar**. The following picture shows how a new *Maple V* session window will look.

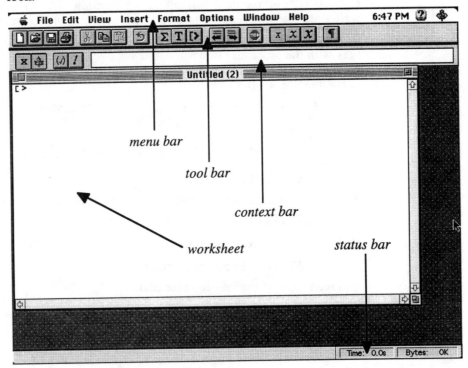

You enter the *Maple* commands in the worksheet area.

For example, when you input:

```
int(x^2, x=0..1);
```

the worksheet passes this expression to the **kernel**, the "computing unit" of *Maple*. It is there that integration is performed and the result of $\frac{1}{3}$ is computed. This result is then passed back to the worksheet and displayed.

Execution Groups

Each worksheet is composed of **groups**. There are different types of groups as you will see later. *Maple* will automatically combine each input and output to form an **execution group**. The groups are indicated by "group brackets" that appear on the left side of the worksheet window, as shown here:

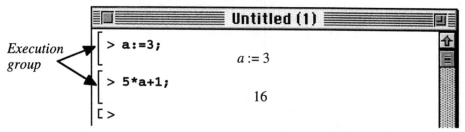

Execution group

Inserting a New Execution Group

After each computation, *Maple* will automatically create a new execution group waiting for your input. However, you can also insert a new group at any space between two existing groups. Let's show you how to do so.

Say you want to add a computation before the input **5*a+1**. There are two ways to do so:

- Method 1: Place the cursor in the output $a := 3$. Press the ⊳ button in the tool bar (see the picture at the beginning of this Appendix). A new execution group will be created before the input **5*a+1**.

- Method 2: Place the cursor in the input **5*a+1**. On the menu bar, choose the **Insert** menu and select **Execution Group → Before Cursor**. A new execution group will be created before the input **5*a+1**.

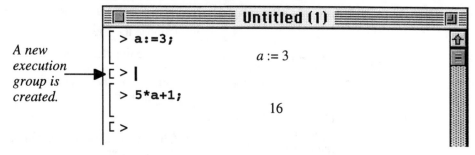

A new execution group is created.

Inserting Text Between Execution Groups

In addition to execution groups for input and output, there are also **text groups**. These allow you to add commentaries, titles, and headings to your work.

Suppose you want to add some commentary before the input **5*a+1**. To do so,

- Following the previous instruction (method 1 or 2), insert a new execution group before the input **5*a+1**.

- Press the **T** button in the tool bar. ("T" stands for text.)

- A text menu bar will show up. Choose the text style and font. You can start typing your commentary

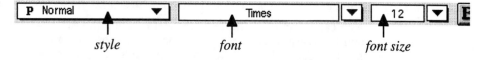

style   *font*   *font size*

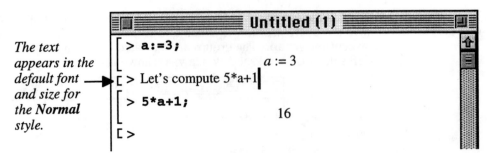

*The text appears in the default font and size for the **Normal** style.*

Using a similar procedure, you can also change the text style to **Title, Author, Bullet Item,** and so on. Each of these styles has its own format. You can experiment with them.

Editing Cells and Text

The following table summarizes the editing procedures you'll use the most when working in a worksheet:

What You Want to Do	*How to Do It*
Delete a group.	• Highlight all the material in the group. • Hit the delete key.
Make a copy of a group in a new location.	• Highlight all the material in the group. • Choose the **Copy** command from the **Edit** menu. • Create a new execution group at the location and use either the ▶ button in the tool bar or the **Execution Group** in the **Insert** menu. • Choose the **Paste** command from the **Edit** menu.
Cut, copy, or paste the text of a group within the same or another group.	• Handle this the same way that you manipulate text in any word processor. (Use the mouse to select, and then use one of the **Cut, Copy,** or **Paste** commands.)
Change the font, size, or style of some (or all) of the text within a group.	• Select the text with the mouse. • Choose the appropriate font, size, and style from the text menu bar.
Change the default font, size, or style of all the groups.	• In the menu bar, choose the **Format → Style** menu. Select the appropriate font, size, and style.

Worksheet Organization

Titles, Sections, and Subsections

A well-organized document has a title, perhaps a subtitle, and is divided into a number of sections. The sections may further be broken into subsections.

You can structure a *Maple* worksheet in much the same way by using the **Title** style and the **Indent** feature. Using these together with the execution groups yields a nicely organized document that's suitable for both reading and printing.

■ **Example.** Suppose we have made the following simple calculus computations in a worksheet:

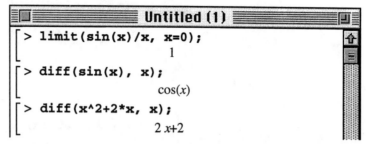

To organize these results, we will add a title named "Math examples." This can be done by choosing the **Execution Group** → **Before Cursor** in the **Insert** menu, clicking the **T** button in the tool bar, and choosing the **Title** style as we mentioned in the last section.

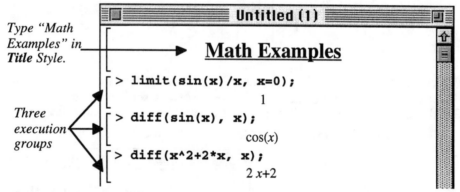

Type "Math Examples" in Title Style.

Three execution groups

Now we want to add a section heading to say that these are from "Calculus."

- First select the three execution groups (see the picture above) by highlighting them.

- In the menu bar, select the **Format** menu and choose **Indent**. A section group is created.

- Move the cursor next to the ▤, and type "Calculus." (See the picture below.)

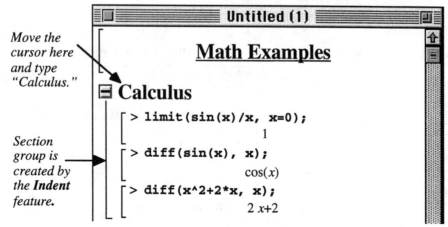

Move the cursor here and type "Calculus."

*Section group is created by the **Indent** feature.*

Similarly, you can add two subsection headings that separate the material into

"Limits" and "Derivatives."

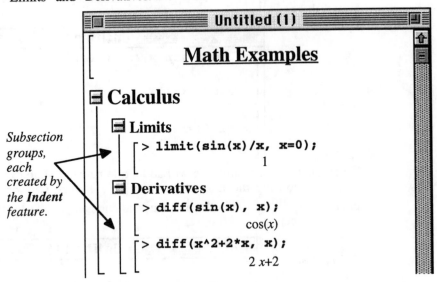

*Subsection groups, each created by the **Indent** feature.*

Open and Closed Sections

All the sections of the previous window are said to be *open*, because you can see their contents. You can *close* a section or subsection so that you only see its heading. This provides a very nice outlining capability, as you'll see.

For example, the "Derivatives" subsection can be closed by clicking its ▤ button.

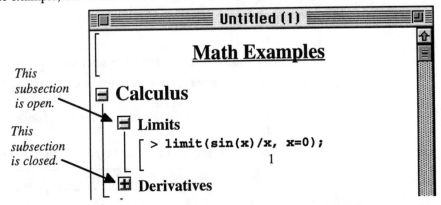

This subsection is open.

This subsection is closed.

If you also close the "Limits" subsection, the window becomes even more compact.

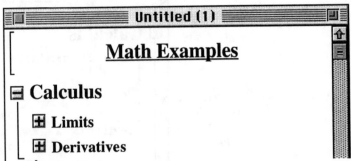

To open the section or subsection, click on the corresponding ⊞ button.

APPENDIX B
Input Shortcuts
(Release 5 or Higher)

Input

In *Maple V Release* 5 (or higher), there are two new shortcuts to enter commands and expressions. They're especially useful for preparing worksheets for printing and publication.

Palettes

In the menu bar, under the **View** menu, you can find the **Palettes** command. There are three types of palettes: **Expression Palettes**, **Matrix Palettes**, and **Symbol Palettes**.

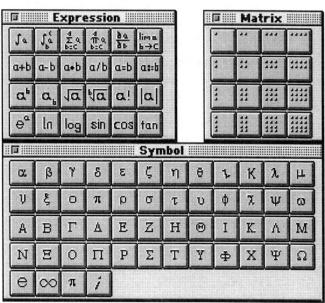

We'll quickly illustrate how they work.

■ **Example**. In Release 4 or earlier, you can only evaluate $\int x^2 \, dx$ by typing:

```
int(x^2, x);
```

In Release 5, you can use the **Expression** palette:

- Open the **Expression** palette by selecting **Palette** → **Expression Palette** under the **View** menu.

- Choose **Math Input** under the **Insert** menu on the menu bar. You will see a highlighted ? next to the *Maple* input prompt >.

- Click on the $\int_a$ button in the **Expression** palette. A new expression is created in your worksheet that looks like:

$$\int ?\, d?$$

- Click on the a^b button. Now you see:

$$\int ?^?\, d?$$

- Type *x*, then *tab*, then *2*, then *tab*, and then *x* again. Every time you press the *tab* key, you move from one box in the expression to another box. Now you see the completed expression:

$$\int x^2\, dx$$

- Hit the enter key to evaluate. You now get the expected result:

$$\frac{x^3}{3}$$

The Context Menu

In Release 5, there is a short cut for you to differentiate, integrate, factor, or evaluate a previous *Maple* output. Consider the result $\frac{x^3}{3}$ we obtained earlier. Suppose we want to differentiate this result.

- Click the output $\dfrac{x^3}{3}$ with the right mouse button (or hold down the option key when clicking for Mac users). A **Context** menu pops up.
- Select **differentiate x**.

Immediately, *Maple* will automatically enter the input and show you the result:

```
R0 := diff(1/3*x^3,x);
```

$$R0 := x^2$$

(The name **R0** is automatically generated by *Maple* for your reference.)

Other Uses of Palettes

Greek Letters

Prior to Release 5 of *Maple V*, you had to spell out the name of a Greek letter to write Greek! You might define a variable named **alpha**, in order to see the Greek character α directly in front of you. In Release 5, however, you can click the α button in the **Symbol** palette; then *Maple* will enter the word **alpha** automatically for you.

For example, you can input an expression that assigns 3 to the variable α by clicking the [α] button, then typing ":=" and "3". *Maple* will automatically type:

```
alpha := 3;
```
$$\alpha := 3$$

Now you can click the [α] button anywhere you'd use the variable α, such as with:

```
4*[α] + 1;
```
$$13$$

This is particularly helpful, say, when you need to enter spherical coordinates. (The output is not shown here, however.)

```
plot3d([cos([θ])sin([φ]),sin([θ])sin([φ]),cos([φ])],
       [θ] = 0..2*[π], [φ] = 0..[π]);
```

Entering a Math Expression in Text

We can use the **Expression** and **Symbol** palettes to enter mathematical expressions into the text of a worksheet. Suppose you are in the text group (please refer to Appendix A for information about text groups). Now, you want to add the comment:

"We need to calculate $\dfrac{\partial}{\partial x} 2xy^2$ to justify the answer."

- Enter the first four words as usual.

- Enter the formula $\dfrac{\partial}{\partial x} 2xy^2$ by choosing the [a a / b b], [a•b], and [aᵇ] buttons and typing the appropriate variables in the equation.

- Press the [T] button in the tool bar to change back to the text mode and continue typing the remaining sentence.

Index

Notes

Notes

Notes

Notes

Notes

Notes

Notes

Notes